Mathematics and Information in the Philosophy of Michel Serres

Michel Serres and Material Futures

Serres was a radically disruptive thinker with respect to the history and practice of philosophy and the future of enquiry. His writings stretch from the sixties of the last century to up-to-the-minute discussions of the present ecological and economic crisis.

This book series will provide discussions both on the various stages and aspects of his own writings, and extend the discussion to developing responses to the questions and problems of and for philosophy, which his writings have brought to the attention of his readers. The return of a concern for mathematical innovation in philosophy is one such area of interest, as indeed is the focus on ecology and climate change.

Series editors

Professor Joanna Hodge of Manchester Metropolitan University, UK
Professor David Webb of University of Staffordshire, UK

Series editorial board

Professor Claire Colebrook, Edwin Erle Spark Professor of English, Penn State University, PA, USA
Professor Steve Connor, Grace Two Professor of English, University of Cambridge, UK
Dr. Diane Morgan, Senior Lecturer, School of Fine Art, History of Art, and Cultural Studies, University of Leeds, UK
Professor Dan Smith, Department of Philosophy, College of Liberal Arts, Purdue University, IL, USA
Professor Iris van der Tuin, Utrecht University, Netherlands
Dr. Chris Watkin, Senior Lecturer, French Studies, Monash, Melbourne, Australia

Other titles in the series

Michel Serres and the Crisis of the Contemporary, edited by Rick Dolphijn

Mathematics and Information in the Philosophy of Michel Serres

Vera Bühlmann

BLOOMSBURY ACADEMIC
LONDON • NEW YORK • OXFORD • NEW DELHI • SYDNEY

BLOOMSBURY ACADEMIC
Bloomsbury Publishing Plc
50 Bedford Square, London, WC1B 3DP, UK
1385 Broadway, New York, NY 10018, USA
29 Earlsfort Terrace, Dublin 2, Ireland

BLOOMSBURY, BLOOMSBURY ACADEMIC and the Diana logo are trademarks of
Bloomsbury Publishing Plc

First published in Great Britain 2020
This paperback edition published in 2021

A catalogue record for this book is available from the British Library.

A catalog record for this book is available from the Library of Congress.

ISBN: HB: 978-1-3500-1976-8
PB: 978-1-3502-5132-8
ePDF: 978-1-3500-1977-5
eBook: 978-1-3500-1975-1

Series: Michel Serres and Material Futures

Typeset by Newgen KnowledgeWorks Pvt. Ltd., Chennai, India

To find out more about our authors and books visit www.bloomsbury.com
and sign up for our newsletters.

For Werner Oechslin, by way of thanks

Contents

Preface

There is noise in the world before we raise our voices there, before the crowd makes its grumbling cries. Lightning writes its forked inclination in the sky; birds trace their flight there, in direction and meaning. There is meaning in space before the meaning that signifies. Taking auguries is believing in a world without men; inaugurating is paying homage to the real as such ... What's incomprehensible is that this meaning, one day, became comprehensible. The physicist is an augur who succeeded.

Michel Serres, *Rome*[1]

Tables of numbers replace tragedy. Chance no longer has a project, but only combinations.

Michel Serres, "Vie, information, deuxième principe"[2]

Michel Serres was a polymath thinker who has written more than forty books, who had been a member of the Académie Française for more than twenty years, who had been a professor in Stanford, California, for many years but whose work has yet, due to reasons that are doubtlessly *no simple* reasons, hardly been received in scholarly discourses. This does not mean that it is not widely read, as the growing number of translations testify. That Serres's work is so little—or so privately, perhaps—engaged with is only consequential, I have heard people say, to what he apparently sought to achieve with his books. They are written in an original style, largely without technical jargon and in a poetic prose; and then again, they treat eminently technical and hugely authoritative subjects like mathematics, religion, history, physics, communication, as well as literature, sculpture, and painting. Serres engaged with these subjects in a somewhat idiosyncratic way—we only seldom find commentary or critical appreciation of the work of other contemporary scholars who engage with similar questions. Serres himself called his discourse one "in Exodus," one that can never take place "at anyone's home terrain." It is a discourse that keeps its distance from epistemologies that are primarily interested in the (subjective) history of the development of ideas. It is as if, where discourses on the history of ideas today place "epistemology" as their organizing principle, Serres positions geometry. His concern is with history in its "objectivity." I read Serres's entire oeuvre as a contribution to our contemporary concerns with the Anthropocene: human life has turned into a natural force that manifests on the geological scales of the earth. The "objectivity" Serres talks about with regard to history is constituted geometrically, and hence it is ideational, but it is entirely committed to an architectonic realism (or a realist classicism, as Chapter 7 will argue) for which the guiding question is: "If it is not a miracle, then can we build it?" Serres relates the role of geometry in the tradition of idealist philosophy with that of physical forces in the materialist tradition. Both, the mechanics of the heavens (and its projected Harmonics, cosmology) as well as that of the earth (and its projected harmonics, technology) are constituted by linking theory and geometry. We will see how this particular gesture of attending to geometry is profoundly informed by a part of the natural disciplines in classical and pre-classical antiquity that has been largely neglected in modern science and philosophy, namely, the branch that studies what antiquity named "meteora": phenomena *above* the surface of the earth (but below the spheres of the heavens) and their connection with phenomena *below* the surface of the earth. Both of the two motivic keys above (the epigraph) help us to unpack the implications of this.

Let us look at the first one first: "There is noise in the world before we raise our voices there … [T]here is meaning in space before the meaning that signifies." Serres is not directly asking the same kind of questions as the ancient meteorologists used to ask—questions like "why is the sea salty," or "how are we to explain the semi-circular shape of rainbows," or "why is there so little hail in higher geographical regions"; but he is deeply committed to their tradition of attending to natural phenomena because with respect to the meteora, it is as if any directive factor of causality (teleology) is being suspended within a cyclical kind of motion and change. The so-called *meteora* phenomena are treated in terms of polar cycles (cycles of evaporation and condensation, the cyclical path of the sun between the equinoxes, all sorts of recycling of matter into different forms). Serres's writing seeks to establish a natural philosophy that continues this tradition, and information theory plays a central role therein. Thought and knowledge in Serres's philosophy are relative to domains that must count as *objective*, and yet these domains manifest in a *heteroclite* manner, with reappearing regularity (globality) but irregularly inflected (locality). What Serres calls "objective" is *multiplicitous*, and cannot be added up smoothly, or be integrated into a larger whole easily. Phenomena in their objectivity need to be articulated, and pronounced. Their objectivity asks to be "bridged" with the objectivity of other phenomena. Serres's notion of objectivity is constituted categorically—it is not a property of classifications; it is what makes classification possible. This is another aspect of the meteora tradition and the study of natural phenomena in their cyclical polarity: it achieved such bridging by operating with what Aristotle called "commensurate universals." Commensurate universals are properties of which one seeks to prove, by demonstration, that they are coextensive with the subjects to which they belong. An eminent example for such properties is the triangle in geometry. A triangle has the sum of its interior angles equal to two right angles, and this property ought to be attributed to all triangles and not, for example, only to isosceles triangles. While the property certainly belongs to the latter as well, it does not belong to them by virtue of being specifically isosceles triangles (what is specific to an isosceles triangle is that it has two equal sides). The important aspect about the concept of commensurate universals is that it asks us to think a *mediate* relation between a given subject and its theoretical formulation *as* a given subject (by means of attributing properties to it). This mediacy constitutes in Serres what he calls the "objective transcendental." The mediacy of this relation can only be *modelled*, but at stake thereby is a notion of a model that *realizes* "objectively," not one that *represents* objectively. The realization at stake in this notion of the transcendental is to be conceived in the cyclically polar terms of the meteora tradition. Models realize objectively only insofar as they make the "realization" as well as the "de-realization" of their objects conceivable. For example, before Aristotle wrote his treaties on meteorology, the discipline was already well established but had actually been in decline, due to the novel theories that attended to change through the very general principle of locomotion (in response to Parmenides' refutations against change), or also from the point of view of ancient atomist theories.[3] From their point of view, mainly the highest level of generality (e.g., chaos, void) attracted scholarly attention, and the rich wealth of the common world meteora phenomena were only of subordinate interest. But while the highest level of generality provides the most authoritative explanations, interest in this level also conflicts with an empirical approach to physics and nature, and hence tends toward idealism. The empirical approach in the meteorological tradition is supported by a maxim of adequacy: a number of hierarchies must count as integral to "the greatest genera," and those hierarchies can be separated by applying a "more or less" scale. This method had motivated a notion of natural order in the terms of *scalae naturae* (natural scales). To a certain degree, Serres affirms this tradition of thinking in terms of scales, yet he complements the method of classification in terms of "more or less" with one that operates in terms of "maxima and minima." This change in method is important, and it is also why I will be using quite abstract terminology at some points. This is not to defend the authority of a particular discourse, but simply to keep this distinction at work. I will refer to such *natural scales* that are subject to Serres's method of "maxima and minima" as "scalarity" rather than as "a principle of scaling," or simply as "scaling." This is to contextualize this method in the meteora discipline of natural science, but also to distinguish it clearly from the notion of scales that work as a representative model in the tradition of *scalae naturae*, insofar as it tried to make classifications where typically there was "naturalized hierarchy" among the different species. The "scales" in Serres's approach are models that "realize," not models that "represent." While no single species crowns the highest steps of the ladder in Serres's philosophy, there is a *greatest genus*. It is what organizes the commensurate universals of his notion of universal nature. This property that belongs to the greatest genus in Serres is that of intelligence, and the subject to which he attributes it is the universe itself. There is an intelligence that is not only coextensive with the universe, Serres maintains, it is also immanent to it. This *natural intelligence* (together with its equally valid, inverse formulation *intelligent nature*) is what lets Serres attend to nature through the lenses of information theory and quantum physics. They bring us the perspective of a manifold universality; namely, that all existing things (animate or inanimate, mineral or meteorite) receive, send, store, and process information. The natural scales in Serres do not facilitate the representation of a natural order, but the transversal passages (the "physical communication") by means of and within such an order. What I will call "scalarities" simply refers to such models that facilitate "passage" (the receiving, sending, storing, and processing of information) within the great genus of such multiplicitous universality. They provide the modules (the one over many relations) that constitute the transcendental objective in Serres's philosophy architectonically. They facilitate the passage of information—quanta of this massive natural intellect that communicates naturally; quanta that from a quantum physics point of view are capable of "traversing" or "communicating" (exchanging charges) between the orbits of various "energy niveaus."

Serres not only affirms that "There is noise in the world before we raise our voices there … [T]here is meaning in space before the meaning that signifies." I chose also another motivic key for this preface, to clarify the relevance of Serres's philosophy for the Anthropocene. It is more concerned with the position of subjectivity rather than objectivity in his thought (although it also illustrates that with respect to the cyclical polarity according to which meteora phenomena can be associated, the two always entail each other). This second motivic key is that "Tables of numbers have replaced tragedy. Chance no longer has a project, but only combinations."[4] With this statement, Serres takes a stance of considerate distance to one particular and ethically important narrative that accompanied twentieth-century industrialization and progressivism, that of the tragic human nature as proposed by psychoanalysis. Sigmund Freud nominated the tragic king, Oedipus, as the common persona that lets us recognize the human in each other. It has been an eminently important idea for twentieth-century social and cultural theory (and political philosophy), to identify with each other through the persona of Oedipus as the embodiment of (tragic) humaneness-in-general. The distance that Serres keeps from this narrative is that the world in which one is a subject, for him, is a world "of objects among objects." Objects are to *model* the central position for collectively subjective identification, which Oedipus provides for psychoanalytic accounts. With this turn to the objective, we ought to respect Oedipus, not only as tragic but *also* as intellectually adventurous—adventurous not in how he seeks to bypass the prophecy, but in how he "actively-statically" seeks to "keep" (maintain) peace with the Sphinx. Serres's twist to the narrative is to direct attention to how both, Oedipus and the Sphinx, actually place their lives in each other's hands (see Chapter 0, "Instead of a Conclusion: The Static Tripod" which cites Serres from *Statues* at some length). This is not to dethrone the idea of affirming a certain kind of castration as a cultural prerequisite for social coexistence; but it reshuffles the cards. It is the background against which Serres affirms, in *Hermaphrodite* (1987), the intellectual "neutralization" of one's sexual body (castration). But for Serres, neutralization works by a dissolution of concentrated power into a distributive plenitude of it. This neutralization reconnects the sexes with a universal genitality rather than introducing a moral abstinence with regard to "giving in" to biological sexuality. There is space for an intellectual kind of eroticism that does not find fulfillment in reproducing the same (thoughts), but in a novel idea of intellectual joy—"lovemaking in the age of the pill" as Serres also calls it.[5] He treats neutrality and castration according to a categorial method of maxima and minima scalarities: not "more or less," or "either/or": "Neutral expresses well enough the inclusion of the excluded middle: neither 'neither one nor the other' nor 'both one and the other'. Castration plays the role of the neutral element, here, for every operation that brings alterity into play."[6] For Serres, as we will see, myth (Oedipus) and algebra (identity statements, equations) are two poles of such scalar polarities which "cycle" the "universal commensurables" of his proposed (anthropocenic) meteora wisdom. In contemporary algebra, the neutral element is what makes it possible to factorize a relation between apparently tautologous terms (an equation) in variously articulable and balanceable ways (by relating to the terms through code). With his twist to the Oedipus plot, Serres proposes to introduce such a level of "coding" also to discourse. With it, discourse cannot only be about "being right"; it needs to take into account also its other pole of a modelled symmetry. There is always a "being left" that accompanies notions of righteousness. Discourse still is about truth; and it is about value, and about usefulness. But it needs to be so on *any* scale. This is how Oedipus is intellectually adventurous; epic without being in the classical sense a hero. The intellectual neutralization at stake involves a likening to, a becoming familiar with what one is not (yet) "naturally" so (here "natural" in the biological sense of "according to one's nature's specific sex").

This has important implications for all philosophies that either privilege difference and alterity, or sameness and identity—in the light of such meteora wisdom, the two need not be regarded as being in binary opposition. According to the notion of modelling that Serres promotes (not models that represent but models that realize in "castrated" manner, if we may say so; a realization that is not the reproduction of the same because it factors in a neutral element) difference and identity can be viewed as poles (relative to cyclical models) that provide for scalar degrees of both to coexist. In all consequentiality, Serres maintains that we need to think about identity in terms of complex *"appartenance"* (multiplicitous relations of belonging-to, belonging-with and belonging-among) in a world of objects among objects, where there is an impersonal, a neutral kind of subjectivity proper to objects. His philosophy, hence, is a positivism (it affirms the progressing levels of abstraction and degrees of power science achieves, and hence history as a power), but one in which oppositions "need to keep their peace" like Oedipus and the Sphinx. Oppositions are poles relative to models that are to realize a great plurality of universal and continuous cycles in which there can be coexistence. Unlike the positivism of Auguste Compte (whom Serres greatly admired), it is a positivism that does not declare spirituality to be a private matter; like rationality, spirituality here is an integral part of the intellect that coextends with the universe (the "one," universal intellect). Serres's view amounts to an ethics according to a paradigm of "positive code" that accepts only "givens" but no (particular) original reference order for how the "givens" come to be "given"; code gives us the "ladders" (*scalae*), but not the endpoints where those ladders straddle (heaven and earth). Meteora refers to phenomena that involve both cosmology (eternal mechanics of the heavens) and "stuff" (substance-matter that knows no change but only continuous transformations among its elements, fire, water, earth, and air). The two mingle in the domain of the meteora, where they are both coefficient with each other. This is also how the meteora domain accommodates change (not only continuous transformations). All the variable ways in how the heavens and earths are coefficient with each other is what this positivism of code (with its polar models, the scalarities) "model" (realize). It is important to say that these models "realize," because it is how Serres's proposed return to the meteora tradition provides a valid philosophical context for the Anthropocene (with its key insight that human beings have turned into a veritable natural force with regard to the economical and ecological balances of the earth). The perspective of his communicational physics allows us to think about how there is discontinuous transversality to these different scalarities. Code facilitates the import and export of values to factor-in together in constellations of complex, multilateral valorization: his positivism accompanies a certain economy of thought, and it naturalizes this economy; but it also pluralizes such economy into natural economies that are *local* and *contracted* across various scales. What Serres's physics of communication, which is at the same time a communication of physics, aspires to respond to is how to think "transit," "traffic" across such heterogenous scales. It is a probabilistic notion of reason, in which chance must be thought to factor in; and it is why chance needs to be considered as "objective."

One must not forget that Michel Serres was, primarily a mathematician. In mathematics, the aim is to formulate something as objectively, clearly, and in as elegantly beautiful a manner as possible. It is very often impatience that makes the texts by Serres somewhat difficult to access, and so dizzyingly fast in the pace at which they proceed. But this impatience is coupled with his style of writing about mathematics (and evidently thinking in terms of mathematics) in a fashion that goes much against the grain of twentieth-century philosophy. I want to suggest that this "strangeness," in how Serres knew, loved, and utilized mathematics to formulate his ideas, came from his interest in this neglected tradition of meteorology. It may be helpful to name and list some of the obstacles that might stand in the way of appreciating Serres's views, such that they can be bracketed and embraced productively. For example, one hardly recognizes in the Leibniz of Serres's thesis *Le système de Leibniz et ses modèles mathématiques* (1968) the same philosopher whom Bertrand Russell has written about. In like manner, when Serres speaks of "la théorie des ensembles" (set theory), one hardly recognizes the discourse on set theory in contemporary philosophy of mathematics. The same goes for his attention to cybernetics, biology, chaos theory, information theory, algorithms, computation, and artificial intelligence. Over the last decades, the clarity and beauty Serres found in the ideas that fertilize those scientific fields has even grown more obstructed by the fact that some of his recognized intellectual guides, like Jacques Monod especially, developed a "reputation" in the specialized communities that make them stand, as influential fathers of novel fields in science (molecular biology), for something quite opposite of what they initially gave rise and credit to. Monod has come to stand, for example, for a branch of doing molecular biology in a crudely deterministic-mechanical manner, and this despite the fact that his initial book, *Chance and Necessity* (1971), was celebrating how biology is just about to begin to attend and explore—quantitatively yet philosophically—the genuine strangeness of its subject matter (i.e., life). Monod's affirmation of stereometry and stereospecificity in biochemistry is capable of and, by Monod, arguably also meant to do entirely without a fixed coordinating reference frame. To unsettle such a point of originality, arguably, is exactly what Monod aimed at with his notion of invariance; the point of origin, if we read Monod with instructions by Serres, is *noisily restless*.[7] The counter reputation he has come to stand for today is, very likely, due to an extraordinarily performative manner of handling chance calculations computationally within a logicist, formalist paradigm of Turing Computing—a paradigm that was neither thought adequate for biology by Turing himself (after his "On Computational Numbers" [1937] he attended to, and was much fascinated by, computation in relation to chemical morphogenesis [1953][8]), nor by Monod or Serres. A similar fate as that of Monod's and Turing's has also happened to George Boole, who today stands for an interpretation of algebra (so-called Boolean algebra) that is decidedly what his book *The Laws of Thought* (1853) was *not* about.[9]

On a less historiographic level, I have found a few dominant ideas regarding contemporary science and philosophy in today's discourses that pose peculiar obstacles for tapping into the rich imaginaries Serres's thinking has to offer. Despite their counterintuitiveness, they are imaginaries that can put us in touch again in novel ways with the very ordinary world in which we live—beyond what Serres once called "the world's reduction to black ink on white paper."[10] With regard to these ideas, the reader will have to grant the credit of placing the dominant theories within brackets as hypothetical, in order to reconsider them anew. What Serres's philosophy teaches us more than anything else perhaps, is that the best way to deal with obstacles is to learn how to embrace them, and not in trying to overcome them. The obstacles I want to mention are the following:

1. Computation cannot be reduced to the dominant paradigm on computability today, which is Turing Computing. Computation is arithmetical, yes, but algebra is more abstract than any one calculus in particular.
2. Mathematics is not a system of rules; while linguistic language would provide the means to make sense of these rules. Mathematics itself behaves like a language. It provides a lexicon of indexical concepts. Of course, it can also provide for systematic frameworks or support structures (in a great plurality of manners and fashions). There is an *objective mentality* at work in mathematics, an anonymous intelligence that is capable of and knows many ways of how to *discern*, make *distinctions* and *relations*, and be *concerned*.
3. In real nature, generative as well as generational, no system is ever closed. Combinatorial games are not opposed to a generative processuality in nature— they are a kind of ship in a bounded space that is noisy (*massively* noisy, hence a bounded space that is "leaking") before there is a humanly imposed significance. The investment in play is a serious, and existential matter.
4. Philosophy does not lend itself well for giving a retrospective legitimization of facts. There is always a prescription that precedes any lawful statement (particularly, to name a striking example, while there is a decision for tradition in Common Law, there is one for systematicity in Civic Law). At the core of philosophy is law and nature, but we ought to refuse to make a categorical distinction between physical and juridical law. Philosophy *anticipates*. (Serres's difficult relation to epistemology—in the paradigms of phenomenology as well as those in so-called analytical philosophy—results from this).
5. Information is a formal, mathematical notion. We have a *quantitative* concept for it today. Consequently, there is an aspect on which information is *not* something different for electro-engineers and linguists, journalists and biologists, physicists and lawyers, or artists. Information, as a mathematical quantity, is categorical (universal). This categorical sense is, arguably, what philosophy today needs to come to terms with.
6. Consistency and exhaustiveness of knowledge about something cannot be integrated together in a nonarbitrary system. Structure and system, as epistemological paradigms, provide for a certain *convertibility* of their objects. They provide for models that are inevitably *reduced* models. We need to let go of the claim to exhaustiveness in order to preserve local, but objective, criteria of consistency.

All of the above-mentioned points require one particular skill: how not to feel at home in one's thoughts; how to exercise thinking in *daring to invert habitual ideas*. By invert

I mean neither to affirm, nor to negate. To invert entails cultivating a neutral stance—and more than anything else, modern mathematics can help us to do so. Mathematics, per definition, remains indifferent to the contents it deals with. This is of course also what has brought mathematics in such close contact with theology in the past. But there is one mathematical idea especially that is key for a veritable modernization of mathematics: that of groups. Groups give us a way of doing mathematics categorially, and thereby they equip us with the necessary means to prevent any (one particular) joining of forces between dogmatic ideologies and mathematical reason. There is a particular maxim that orientates Serres's interest in science and technology, namely, that philosophy needs to maintain (actively and precariously so) compatibility with science. If we want to deal responsibly with the power of contemporary technology, we find ourselves "condemned today" (as Serres puts it) "to become intelligent and inventive." Dealing responsibly means in a manner so that neither does science dominate social orders, nor the other way around (that the customs and traditions of social orders dominate science).

But this pronounced "condemnation" to become inventive and intellectually adventurous is actually an invitation. It is an invitation for everyone to start caring about the vulnerability of the earth, the world, the incredible diversity of life forms, biological as well as cultural ones. It is an invitation to be amazed, and to feel awe at the miraculous—but not unreasonable—*objective* reality in which we find ourselves. The universe is *secretive*, literally so: It "sets itself apart," as the Latin word *secretus* reports the tradition of Proto-Indo–European heritage, from the root *krei-*, for "to sieve," thus "to discriminate, distinguish." When "secrets" play an important role in Serres oeuvre, it is not in the sense of enigma, of a "statement which conceals a hidden meaning or known thing under obscure words or forms." Serres does exactly *not* seek to speak in obscure sayings, to speak in riddles. This is why he insists that the universe itself is "secretive": it sets itself apart, as it literally *expands*, at least according to the models of contemporary astrophysics. Contemporary science may have a good idea about the age of the universe, but this does not make its *fact* any less miraculous. In fact, by attributing a secretive character to the universe itself, as a property of its nature, Serres actively refuses to accept a mysticism around a kind of intelligence that would be "artificial" (as opposed to "human"). Rather, we need to think from an *objective* point of view—with regard to an "objective sense of intuition" as he puts it—about the conditions of observations. These conditions need to apply to the anonymous (impersonal) reality of experience, as well as the fact that there is "mens," "mentality" in experience (it is possible to learn from experience). Such natural "mentality" is the mentality mathematics can grow conscious of. It is the mentality embodied in all things that exist, also in objects, techniques, and instruments. It is the natural mentality of an intellect that coextends to and is immanent within the universe. This is something we can maintain scientifically (quantitatively treatable, rather than merely by animistic or spiritualistic projections) only today, with the mathematical notion of information in information theory. But regardless of our novel capacity of acknowledging it scientifically, Serres relativizes any claims to superiority, *nature has always been "intelligent."*

Now a preliminary word perhaps on Serres's relationship to history within such a conception of universal nature. We have been overly preoccupied with personal, individual subjectivity, on the one hand, and timelessly homogenous, generic, objectivity on the other. It is in this spirit that many of Serres's books engage with the history of ideas, and with the history of science. But when engaging with his work, it is crucial to acknowledge that he is not writing a kind of _counter-history_ to dominant approaches. Rather, there is a sense that every present is—radically, materially so—"con-tingent" to (literally "in touch with") every other present. History and contemporaneity are brought into communication in Serres's "anthropocenic meteora approach" to nature in its universality. What he is proposing is to train our thinking in a retrograde kind of movement that interiorizes the said modernization in mathematics; hence the importance of his view that history applies to mathematical ideas themselves, objectively. This is what he calls "mathematical anamneses" (in the plural). _Retrograde_ literally means a thinking that takes one step back, a thinking that does not _immediately_ seek to proceed, to judge, act, or be critical. To move in a retrograde manner is to move away from the object of interest _as a means to get in touch with it_—with greater subtlety, with more finesse. Every present "banks" on an abundant past, as I will call it in Chapter 7. There is a relation hence between such mathematical anamneses, technique, and sophistication (not only sophistry). In order to acquire such perceptiveness, to obtain such skill, it is necessary to affirm a radical kind of solitude as the condition of possibility to find commonality by developing, sharing, and culturally interiorizing novel social imaginaries that are adequate for a contemporary human condition that is not without hope, that does not give in to misology (the loss of esteem for reason and argument).

Lastly, perhaps, it is important to stress that with these ambitions, themes, and interests, Michel Serres has not been so solitary as this preface now might have made it seem. What Serres develops as _retrograde kind of positivist progress_ articulates an idea that, for example, has doubtlessly also troubled Gilles Deleuze with his interest in what he called _vice-diction_ (in _Difference and Repetition_, 1968), or, together with Félix Guattari, their rhizomatic method and the geophysical stratification of history (in _Milles Plateaux_, 1980), or their idea of "conceptual personae" (in _What is Philosophy_, 1991). Michel Foucault, who could never decide whether he wanted to think of himself as a historian or as a philosopher, was concerned with similar questions when he discoupled an architectonic vector from a genealogical one, in his early book on methods, the _Archaeology of Knowledge_ (1969). In a different manner, Jacques Derrida also evidently found his own paths with regard to inventing a method (a formal and critical way of reasoning) that would neither get caught up in a circularity without direction, nor be committed to a particular path of progressive movement along a line. His _White Mythology_ from 1974, as well as Roland Barthes essay "Le Degré zéro de l'écriture" (1953) obviously ponder the same transformation of writing as Serres did with his "white concepts" in _L'Incandescence_ (2003). Paul Ricoeur, Maurice Blanchot, and Rene Girard along with many others who could be mentioned. How to survive with humane dignity between nonsense (absurdity, without direction) and paradox (dogma, uncriticizable, established by institutional power alone) has been a common struggle throughout the twentieth century.

And yet there is a strange kind of silence with regard to Michel Serres's ideas. This book is meant as a wide and open introduction to his thinking. A note perhaps on why

I chose to write this book in the way that I did, as a more obvious way to raise further academic interest in his work might well have been to begin a play of positions between his own articulations and those of others, with which more people are already familiar. But this is what I decidedly did not want to do: namely, to seek a broader reception and engagement with Serres's work by writing this introduction as "a Deleuzian reading" of Serres, as "a Whiteheadian," "a Derridean," or "Foucauldian," "a Kantian" or any other kind of comparative discussion. This book is to provide openings, not closures. I have been thinking a lot about what it is that I am doing here. The image that sits best is that of what in French is called the work of "une ouvrière, un ouvrier"—a worker-artisan interested in industriousness in a manner that only the French language conveys: not as a worker who strives to *open up* and *lay out* an object she seeks to reproduce, but one that builds models, each one inevitably reduced but with a peculiar finesse, of how something *may lend itself for openings*. Such "une ouvrière" must, together with her models, provide *instruments* for exercising oneself in building such models. This is what I have tried to do throughout my book. The key idea in Serres's philosophy is indeed that of an impersonal cogito, a cogito that thinks objectively in a world of objects among objects, but that does not seek to dominate, but rather the reverse; a cogito of an impersonal kind of subjectivity, that seeks instruction *from* its object. To say it plainly—for quite a few (if not all) of the lines of thinking which I develop throughout these chapters, there is not really one particular direction in which the reader might follow me. It feels like I was involved in a kind of intellectual dance, developing the ideas, following them just a little bit further, and calibrating their adequacy from step to step. I have taken the liberty of leading the movements at certain times, while at others (I would like to think) I have let myself be led entirely by directions from the texts I have been working with. It has been written with great joy.

Accordingly, I have kept the bibliography to a minimum. I have only listed those books that had, explicitly, a guiding influence in this adventure. It goes without saying that a more exhaustive list of influences would quickly turn countless, as I am strongly convinced that even though thinking is a somewhat solitary business, one never thinks alone. Secondly, and for the same reasons, I have worked with a large amount of citations. The path I draw throughout Serres texts, and throughout the references he works with, are by no means intended to give an exhaustive discussion of his oeuvre. They are colored by my own journey on which I have, in the past ten years and through engaging with Serres's writings, learnt to appreciate the wit, the cunning, the beauty, and the amazing mysteriousness that coexists in technics as much as in nature, and of course in culture. It has been immensely inspiring with respect to imagining a basic kind of literacy in how to think about coding as an architectonic "alloy-praxis" that glues letters with numbers, physics with information, mathematics with language— and that resonates with the tempers of a "civic kind of anarchism" rather than with tempers tuned to militaristic (or militarized) information technology; be it Intelligence Services or Hacker Cultures or Corporate Industry-led Education Centers and their Coding Bootcamps (as for example 42, a private and radically industry-led school calls its tuition free program, in which young people are trained without content curriculum and in operational skills only, as if for an army of mercenaries). My hope is that the reader closes this book with a buzzing mind full of questions and ideas for things to investigate and check, to look up and carry on, in whichever direction. For there is an energy for change and a transcending movement proper to any form of literacy, and in this sense coding literacy too aims at that same goal which Louis Hjelmslev had formulated in his *Prolegomena to a Theory of Language* (1953):

humanitas et universitas.

Acknowledgments

Lastly, what remains is to thank the people who have accompanied me on this journey. First of all, I would like to extend my sincere gratitude to Joanna Hodge and David Webb, the series editors of this exciting new book series, and also the key and critical readers of my book. My thanks also go to Frankie Mace and Lisa Goodrum, editors at Bloomsbury, with whom it has been a pleasure to collaborate throughout this project. Furthermore, I am indebted to the careful and invaluable work of my two copy editors, Michael Robert Doyle and Bill Ross. Michael Doyle has helped me through several early versions of this book with his acute sensibility for words, and in helping me to translate from French to English—both languages in which I am not myself a native speaker; William Ross has helped me to finalize my manuscript and bring it into a form that has gained much greater clarity and flow than I could myself have provided. With regard to my personal cosmos of inspiration and exchange, it is naturally not possible to mention everyone who ought to be. Quite simply because even though writing is quite a solitary exercise, there is at the same time also a profound sense in which it is precisely in writing and reading that we are never alone. I would like to specifically mention Ludger Hovestadt, Elias Zafiris, Klaus Wassermann, Iris van der Tuin, Felicity Colman, Gregg Lambert, Rosi Braidotti, Marcel Hénaff, as well as the Digital Gnomonics group of doctoral students and postdoctoral researchers at the Chair for Digital Architectonics at the Department for Architecture, Institute for Information Technology in Architecture, Swiss Federal Institute of Technology ETH in Zurich, Switzerland, and at the Department for Architecture Theory and Philosophy of Technics ATTP, at the Faculty for Architecture and Spatial Planning, Institute for Architectural Sciences, Vienna University of Technology TU, Austria; here especially Diana Alvarez-Marin, Mihye An, Cris Argüelles, Pierre Cutellic, Michael R. Doyle, Kristian Faschingeder, Georg Fassl, Nikola Marinčić, Jorge Orozco, Gerda Palmetshofer, Poltak Pandjaitan, Martin Ritzinger, Miro Roman, Karla Saldana, Adil Muhammad Bokhari Syed, Admir Selimovic, Riccardo M. Villa, and Mohamed Zaghloul, with whom I have had the immense challenge and pleasure, in our joint research seminar since 2011, of working closely through many of Michel Serres's books while seeking to develop and enrich gradually our particular *imaginaries* of a "Digital Gnomonics" and an "Architecture in the Meteora," for how to find a novel understanding of human intellectuality in co-existence with artificial intelligence technics.

Introduction

A substitution takes place in which science eradicates language—this explains our time.

Michel Serres, *The Five Senses*[1]

Today we have reason to hope for what has, until now, been only an illegitimate desire: "that we might be able to understand in one and the same stroke the Greek miracle of mathematics and the fantastic blossom of Greek myth,"[2] as Michel Serres maintains. It is only now that we can with legitimate desire hope for such a new classicism; this is because today, we can attend to the "Greek miracle" within registers in which Greek mathematics and myth themselves are but cases. This is because today's science knows how to *date* its objects in a new way. History extends now to all that there is; all the scientific disciplines work together in attributing *historical dates* to nature, from the Big Bang (some fifteen billion years ago) to the appearance of life, or that of particular life forms on earth (as well as their extinction). But importantly, as Serres maintains in his 2016 book *Darwin, Bonaparte et le Samaritain, Une philosophie de l'histoire*, such historical dating equally applies now to phenomena that have hitherto been called (chauvinistically so) "prehistoric" (like cultures that have never developed a written alphabet).[3] We need to understand Serres's interest in such dating before the background of his earlier work on the relation between mathematics and myth. In an early essay entitled "Structure et importation: des mathématiques aux mythes," Serres envisaged the implications of what he saw at work in the novel paradigm of extending history to all that there is. He wrote that if we can endow the concept of structure "with a normative, a cathartic and a purgative"[4] definition, then "our time appears capable of reconciling truth and sense."[5] Might there really be a way of thinking in which one would not always find oneself belonging to—or having to take sides—between the sobering coldness of rational reason versus the exciting and consumptive warmth of the dark depth proper to all things that (ultimately) truly matter? It is right for both philosophy and science, according to Serres, to grant this other, this "Dionysian world"[6] of "impenetrable, dense, dark sense in which the human soul, its affectivity and its fate are true" a central place. Are we not, with regard to knowledge per se, always also "dealing with reality, and with what it means to be human, within human time and human sufferings"?[7] It is possible to think about this "on a universal scale," he maintains.[8] Of course it ultimately remains mysterious *that* there is thinking, knowing, feeling, and being here in this world. But are these deep meanings, these dark senses, only "symbols of history," he asks. Serres's answer is negative, "Are they not also, in their ultimate form, in their final determination, meaningful models of transparent structures, of an order of knowing, of intellect and of science?"[9] Serres proposes nothing less than an outlook toward a thinking about intellection according to which rationalization does not ultimately amount (in the calculative sense of Enlightenment thought) to a demystification of life. But does it also prevent us from violently giving in to what Henri Bergson, in *Les Deux Sources de la Morale et de la Religion* (*The Two Sources of Morality and of Religion* [1932]) identified as the "essential function" of society, namely, being "a machine to make gods"?[10]

Serres was a trained mathematician. What triggered his interest in philosophy, as he told us,[11] was a split between the worlds of ideas in which common people lived and those in which the people informed about the sciences lived. There is here a bifurcation at work that triggers the discrediting of an intuitive common sense vis-à-vis expert knowledge. Serres's entire oeuvre is dedicated to finding a kind of understanding, a *form of intellection, perhaps*, that is capable of granting a shared world of ideas between science and culture. It is crucial for the possibility of such a world to emerge and such a sense of intuition to prosper, as I will argue, to accept mathematics as a "silent" kind of language. For Serres, as we will see, structure was not the mysterious key that would open all doors. Structure is a key, but not a mysterious one; it is formal through and through.[12] Rather than mystifying structure, it is the world to which he is ready to grant a certain mysteriousness. In this century, Serres maintains, we have already seen several revolutions in the *basic conceptions* with regard to the sciences, and "moreover, further ones are announcing themselves." They will "unhinge the theoretical universe just as suddenly and just as much as, with the slowness due to their inertia, they will unhinge the world of praxis and the ensembles of technics."[13] With regard to the mysteriousness of the real world, Serres maintains, structure "is a methodical, clear, well defined and elucidating concept"[14] that is capable of putting us in touch with the world's reality. Structure as a methodical concept is not estranging us from the real world. It is true that "We don't have the same dreams anymore, we no longer think nor write as our direct ancestors did"[15]—but this is not a sign of estrangement from the world, for Serres. Rather, it is a sign of impatience with a certain "active cultivation of ignorance" in views that were once well reasoned, but are evidently no longer adequate.

The main line that will be developed throughout the book is that mathematics in Serres's philosophy provides a lexicon of operative concepts, rather than foundations or supporting structures for science. At the core of this endeavor, to reconcile truth and sense, is the problem of ideation. We must reconnect with abstraction a capacity to provide for generosity, rather than seeing in it only a means of reduction, he maintains. The operative concepts of the mathematical lexicon trigger models that must count as "reduced," but not because they would draw a poor representation of the phenomena they "model." Models don't "represent" at all, rather they "realize." Mathematics is the inexhaustible source of a generosity from which such "reduced models" (that are capable of realizing particular phenomena) can draw. Like this, they are the words in a lexicon of mathematics through which we can learn to "sound silently" (to hear and to formulate objectively) how nature speaks.

Mathematics and information are crucial for understanding how the philosophy of Serres at large is an attempt toward a "general treatise on sculpture."[16] It is important to grasp Serres's ambition to formulate such a general treatise as a philosophical treatise in the history of religion, not one in the field of aesthetics. This is because, as he writes, "the itinerary of aesthetics brings us to the feet of statues." But encountering statues from this direction (the direction of aesthetics), "they sit in session or enthroned most often at the far end of temples where they are adored by idolaters who are scorned as superstitious." Contrary to the itinerary of aesthetics, that of the history of religion "meets with and highlights this term of abuse [aesthetics], in itself remarkable and so akin to the statue that it contributes to breaking or overturning."[17] According to the kind of general treatise Serres had in mind (and which the lexicon of mathematical words is to constitute), "sculpture bears ancient witness to the anthropological genesis of experience in general. It carves, drills, and fashions. Rodin is right: gate is the true name of the sculptor's ark."[18] The mathematical concept of structure is crucial for the conception of mathematics as such a lexicon. If applied to experience, mathematical structures provide formal keys for opening such gates: "Thus space and time open up through some gate that yawns or gapes open onto what language calls by the same word: experience. An expert gate, the same term, that is to say, open onto an exterior. The gate is a kind of pass. The world and life lead to a threshold that bars an elsewhere."[19]

Such an understanding of abstraction (one according to a general treatise of sculpture that follows the itinerary of religion rather than aesthetics) hinges in essence on a different understanding of history. Taught by the school of the French philosophy of a history of ideas—by Georges Canguilhelm and Gaston Bachelard among others— Serres shared their concern for making history the object of rigorous science. But unlike them, Serres maintains that a scientific approach to understanding history is possible through mathematics and communication (information theory), not through linguistics and logics. If we are right to claim that there is a *history of ideas*—this is how he formulates his interest—then we must also deal with the inverse of this assumption, namely, that there is an *idea of history*. And we ought to assume between the two stances an inchoative relation of *reciprocating dynamic* as well as *systematic implication*, rather than regarding one as a function of the other. It must be possible to excavate an empirical field for the historians' work—a field where the historian can be critical, one where there are methods of reconnecting a direct chain between a phenomenon and its source.[20] Such empirical fields can be established with the help of structure as a "methodically clear, distinct, and elucidating concept"[21] applied to the mathematical notion of information, not to a linguistic understanding of language in terms of signs. We can call such empirical fields for the historians' work (where the history of ideas is investigated as much as the idea of history itself) the field of a *communicational physics*. The mathematical notion of information does not *signify* the quantity it captures; it *indexes* it. We miss the very character of information when we try to relate it to the passive representation of sense. Information affords an excavation of signification, a remembering kind of understanding in *how* signification makes sense: "The technical image of a black box does not signify anything different from what the word 'empiricism' says: it's a matter of drilling an aperture to reach the inside. Experience: a hole towards the outside; empiricism: a window into the interior. In sum, openings onto another place."[22] Information, with its indexical capturing, works like a sieve, a filter: invariance and variation are not mutually exclusive with regard to the statuesque sculpting of reduced models by code. It is thanks to such a *leaky form* (or rather, *spectral* form) that the mathematical theory of information can facilitate a thinking capable of elucidating culturally significant contents through science (rationalizing), and scientifically significant contents through culture (realizing). With this in mind, we can acknowledge how Serres maintains that,

> The simple and pure forms are not that simple nor that pure; they are no longer things of which we have, in our theoretical insight, exhaustive knowledge, things that are assumedly transparent without any remainder. Instead they are infinitely entangled, objective-theoretical unknowns, tremendous virtual noemata like the stones and the objects of the world, like our masonry and our artifacts [objets ouvrés]. Form bears beneath its form transfinite nuclei of knowledge, with regard to which we must doubt that history in its totality will be sufficient for exhausting them, nuclei of knowledge which are profoundly inaccessible like indelible marks. Mathematical realism is weighed down and takes on that old compactness which had dissolved beneath the Platonic sun. Pure or abstract idealities cast shadows once more, they are themselves full of shadows, they are turning black again like the pyramid. Today's mathematics unfolds, notwithstanding its maximal abstractness and purity, within a lexicon which results, partially, from technology.[23]

Mathematics does not provide support or foundations; rather, it unfolds within a *lexicon* of silent and *operative* concepts. I attempt to characterize this thinking as a *quantum literacy* (Chapter 2), through looking at how code is at play within a mathematical notion of information. What preoccupied Serres's thinking from the first of his books onward was in figuring out how to think about "communication." Information theory shows us, Serres acknowledges, how inadequate our inherited views are that reserve language, interpretation, and "softness" for the human sciences, and mathematics, necessity, and "hardness" for the natural sciences. Mathematical communication accommodates all sciences in a common, universal, "white house" (*la maison blanche*). All things are engaged in the fourfold activity of sending, receiving, stocking, and dealing with information—a universal kind of activity that articulates the two categories of Serres's physics of communication (which is likewise a communicational physics), namely, "hardness" (*energy at great scales* [thermodynamics, entropy]) and "softness" (*energy at small scales* [information theory, negentropy]). Serres proposed a philosophy of what he calls "the transcendental objective" and that accompanied his notion of a communicational physics. Categories, in such a philosophy, are *scalar variants* of the two universal categories—softness and hardness. We need to bear this in mind as we go on. It means that basic notions like quantity, quality, and locality ought to be considered in terms of how they order energies at smaller as well as larger scales. I will try to develop Serres's idea that these scalar variants behave somewhat like musical instruments: through them, nature speaks in many tongues. The articulations of Serres's communicational physics are articulated according to a particular kind of "temporal," "tempered" or also "timely" wisdom proper to phenomena subsumed by the ancients as "meteora." The wisdom at work in "reasoning" those phenomena manifests as the integral summation of all measurable durations (there is a particular relation between time and weather that I will elaborate in Chapter 3). With the help of such wisdom, we can think of the inverse of a natural history of ideas, that is, we can think *the idea of history*; that is because as the integral summation of all measurable durations, the wisdom of the meteora provides something like an inverse of the idea of the *nature of number*. The wisdom of the meteora can be conceived as *the numericalness of nature*. This amounts to the maxim that if we think about counting time, we need to temporalize counting in turn. Nature is countable, but countability too must be considered natural. Code provides the operators of such convertibility. This idea is developed in Chapters 3 and 4, "Chronopedia I: Counting Time" and "Chronopedia II: Treasuring Time," as well as Chapter 6, "The Incandescent Paraclete: Tables of Plenty." The orders of such articulation—its "metaphysics"—are scalar orders (orders in terms of *scalae naturae*), but its notion of order is conceived as a model that realizes, not as a model that represents. This is how we can find in Serres's philosophy a proposal of how to reconnect with this old tradition of organizing rational thought in terms of gradual processes that can accommodate different hierarchies subject to one universal nature, but without imposing a particular hierarchy as representative. Through attending to how time is counted (as measurable durations), Serres proposes a method of maxima and minima, where Aristotle (for example) proceeded in his gradual scheme according to a method of more or less. I call such a manner of organizing thought "architectonic." The hierarchies themselves can be thought of formally with the mathematical concept of structure, this is why they constitute what Serres calls his "transcendental objective." The hierarchies facilitate the "passage" of information across the various scales of "hardness" and "softness." The orders at stake thereby unfold in a great variety of "instrumental scales" (models that "realize"). And like musical instruments too, those scalar "instruments" indeed have a proper "temper"; just as a violin has a different one from a flute, the silent words in the lexicon of mathematics have individual "tempers" too. The great variety of scales can provide for harmoniousness that does *not* depend upon an integration into one exhaustive and pure whole: to Serres, nature speaks in many tongues, often noisily so. Like the weather, nature speaks with temper, cruel, stormy, tender, soft, violent or caressing tones. Here Serres's thinking takes a distant stance from that of Leibniz, who remains doubtlessly one of Serres's greatest influences for these ideas. But for Serres, we need to think of "concord" (harmony) in its irreducible relation to "discord" (violence, noisiness). It was simplistic to merely attribute reversibility to spatial configuration and irreversibility to temporal processes, Serres maintains. Natural things are marked by a material notion of passing time, by ageing. Things are marked by both, reversibility and irreversibility "at the same time," on various scales. A physics of mathematical communication challenges philosophy to welcome the concept of "mass" as a third in the *fundamental* order of physics, next to those of space and time. This is where in such a natural philosophy that sculpture steps in, where classically it was the place of epistemology: "Sculpture, hard, like music, soft, precedes language, the one in its own order and the other in the order of things; the one participates in the little energies, the other in the high ones."[24] Mathematics is a silent language with inexhaustible, unfathomable stocks of meaning. Its language does make sense, but "sense" means not only significance but also direction. Nature is directional, but it is so in many ways and on many scales—the language game of invariance and teleonomy, which assumes an objective notion of chance as the scattering-operator that renders directionality indetermined, will equip us for addressing such an understanding of sense. We come back to this later (Chapter 2 "Quantum Literacy").

It is an unusual idea that mathematics is to be accepted as the medium of transmission in communication. Many philosophers and scientists alike, interested in intuition and empiricism as their methodical stance, would see in the "powerful abstractness" of mathematics much rather the cause for the emerging distrust in intuition and in a commonsense language. Why? First of all, the worlds of ideas in mathematics are *many*. It has been a core interest since the beginning of the last century to unify mathematics into one corpus that rests on firm (rigorous or exact) foundations. But there are different branches of mathematics that can be credited to be of the kind that would make such unification possible: "In these days the angel of topology and the devil of abstract algebra fight for the soul of every individual discipline of mathematics," as Herman Weyl famously put it.[25] With regard to the question of intuition, this may seem to be a storm in a teacup, since any one of these branches is so abstract that most people cannot think of a phenomenal reality of experience where these ideas would make any sense. But this is exactly what Serres opposes when he affirms that "mathematics unfolds, notwithstanding its maximal abstractness and purity, within a lexicon which results, partially, from technology."[26] The lack of imagination of a phenomenal reality of experience results from a particular *disinterest* in the technologies, and the technical objects that disposition and characterize our everyday life. Does it not have a strange appeal when we hear a chemist and electronics engineer write today: "Semiconductors ought to range as an example of human achievement alongside the Beethoven Symphonies, Concord, Impressionism, medieval cathedrals and Burgundy wines and we should be equally proud of it."[27] Why is it, indeed, that there is a feeling of oddity accompanying this demand? Of course, this lack of appreciating our current form of technics is due, partially, to its abstractness and to the degree of expertise it seems to require. But has this not been the case for *any* of the above-mentioned artifacts we are all holding as precious and dear? An understanding of how semiconductor electronics works, by what it is conditioned, and to which ends we might perhaps, at one point, be able to cultivate it, this holds a promise of no lesser awe-inspiring enjoyment than Ludwig van Beethoven's symphonies, the Concord, impressionism, medieval cathedrals, and Burgundy wines. At least for the chemist whom we are citing here: "I only hope that my attempt to explain something of its appeal will help the layman to obtain the same kind of enjoyment from an understanding of semiconductor electronics that he or she might experience in contemplation of any of these."[28] Serres probably could not agree more with our chemist on this. Chapter 7, "Sophistication and Anamnesis: Remembering an Abundant Past," is written as an exercise in learning to achieve such enjoyment.

This example illustrates well that such discrediting of the abstract artifacts (which counteracts the emergence of a more adequate common sense of intuition) is one due to illiteracy with regard to technical objects. While artifacts have always also meant (and had been "rationalized" culturally) to be miniatures of something "whole" (beauty), something "priceless," and "cosmic," today we don't seek to reason and rationalize this dimension of artifacts. This is so much so that we should perhaps best speak of such illiteracy as the inducer of an "*existential* kind of *distrust*." To begin speaking of existential distrust might sound like a contradiction in terms, but this is precisely why I want to suggest it. We tend to think that it is quintessential for existentialist thought that it has rid itself of any notion of belief. But the kind of distrust in technology as providing, in part, the concept of a lexicon where mathematics is unfolding, only shows that belief can hardly be negated—at least not without mobilizing, in turn, some sort of misologist (from the Greek *misologia*, for hatred of argument, reason; the revulsion or distrust of logical debate) enthusiasm for this negation. Just such a kind of enthusiasm reveals itself toward the kind of physics that characterizes our daily lives more than anything else perhaps, because it brought us the mastering of electricity, and provides for all our electronic devices and infrastructures, namely, quantum physics. In a recent interview with Martin Legros and Sven Ortoli, Serres remembers the enthusiasm of his colleagues for a kind of political philosophy that, he says, knew very little about the sciences—discarding quantum mechanics as a whole as bourgeois science in the interests of claiming an honesty and modesty proper to manual work alone, by opposing work (praxis) to intellection (theory).[29] In that interview Serres recollects from his formative years in Paris, a lecture by Louis de Broglie, a Nobel Prize laureate, on his hypothesis about the wavelike character of electrons. The subject was the uncertainty principle postulated by Werner von Heisenberg. Serres's peers from philosophy, so he says, knew very little about those developments in mathematics that have had such an influence on him.[30] To them, passionate only about political activism, what de Broglie was talking about was outrageous:

> The others kept aggressively disturbing the conference, because to them, Marxist determinists as they were, de Broglie was, with Heisenberg, one of the apostles of indeterminism. He was an enemy of the class, a reactionary that needed to be wiped out! I saw de Broglie leave sheepishly, protected by two or three persons, amidst the booing of the students, led by Althusser.[31]

Experiences like these must have been formative for Serres's thinking. He was troubled by the fresh kind of ignorance—a sheer willful illiteracy!—vis-à-vis the developments in twentieth-century science. While he perceived the advent of information and communication technology as a major game changer for the relations between science and societies, and set out to anticipate its impact, most of his intellectual peers were preconcerned with advancing on matters of production rather than attending to matters of communication in their own terms. They aimed at making sense of the novel developments within the framework of nineteenth-century industrialization. This brand of political philosophy claimed that philosophy can only truly be political, and hence in the interests of the public, if it dissociates itself from the sciences. But does this not amount to orthodoxy and ideology? It claims that philosophy can be "practical" (rather than "intellectual") if it only dissociates itself *from* the very practice that drives civic developments (namely, the kind of science at stake here: in quantum physics, electronics engineering, microbiology, etc.). To Serres, this is not political philosophy but the very betrayal of philosophy.[32]

Let us pick up our strand of thought where we left off. Semiconductors in particular are the very material embodiment (and at the same time a very abstract explication) of how quantum mechanics challenges us to reconsider philosophically classical ideas of causality, measurement, calculability, ontological, or modal conceptions of possibility and necessity. It is also the core idea worked through, philosophically, in Serres's book *The Parasite* (1980). We can only adequately understand the irreducibly triadic relation between sender, receiver, and interceptor which he insists on in this book, if we acknowledge as the "silent word" through which he studies communication, namely, the mathematical formula for "semiconduction" in discourse, that constitutes philosophy according to "a language of many portals":

> If the "guest" is a tax-collector, in the broadest sense, I consider him to be a parasite in the political sense, in that a human group is organized with one-way relations, where one eats the other and where the second cannot benefit at all from the first. The exchange is neither principal nor original nor fundamental; I do not know how to put it: the relation denoted by a single arrow is irreversible, just takes its place in the world. Man is a louse for other men. Thus man is a host for other men. The flow goes one way, never the other. I call this semiconduction, this valve, this single arrow, this relation without a reversal of direction, "parasitic." ... The system constructed here beginning with a production, temporarily placed in a black box, is parasitic a cascade. But the cascade orders knowledge itself, of man and of life, making us change our terminology without changing the subject. It is an interesting circuit which we shall follow in order to understand one thing, various landscapes, several epistemologies. Maybe polyphony is in order. I call the language of many portals "philosophical."[33]

Serres himself was very much what one would call a public figure, but not in the sense of what the existentialist tradition has celebrated as a public intellectual. Vis-à-vis this tradition, Serres's public persona may well appear conservative; he was a member of the Académie Française. But his philosophy is not what one would call today "academical." It is not geared toward founding a secluded place to withdraw into and concentrate. It tries to live up to nature in the open, to nature speaking in many tongues. It seeks to find concepts capable of grasping phenomena in different domains all at once, without reducing the domains to one hierarchical, deductive order. He did not want to explain things in the sense of demystifying them from the bottom up. He did not try to sort things in their purity from things as they appear. He did not lecture the uninformed on the right way of orienting their thinking. What is at stake with his overall ambition, namely, to develop how our time can reconcile truth and sense, was always to find a sense of intuition that is nothing more, but also nothing less, than *trustworthy*: "If philosophy doesn't have to dominate science or become its slave or handmaiden," he formulates such trustworthiness, "it must at least maintain compatibility with it."[34]

To Serres, it was just such a *compatibility* that mathematics, the provider of a lexicon of operative concepts (theoretical models and instruments for an empirical praxis) can establish. As with every literacy, it takes time and exercise until playing with its instruments can be interiorized; until one develops a sense of what is possible with them, so that one can begin to work intuitively. Compatibility in mathematical terms exposes *the conditions* for something to be considered reasonably as trustworthy (not as true or false). The ethical maxim therefore is to seek neither domination nor slavery to science and its developments. Serres's project, with regard to developing *a sense of intuition* that preserves compatibility, was hence a transcendentalist project. But unlike Immanuel Kant, Serres does not situate intuition in the faculties of a generic subjectivity. He found it in those of a naturally contracted objectivity. His transcendentalism too is to make space for criticality. However for Serres, mathematics alone was capable of providing and maintaining that space. This was different for Kant, for whom this space was to be one of *criticism* (not *criticality*), and it was to be referenced in an immediate manner by the juridical. Kant's transcendentalism was a political not a natural philosophy. The words of a mathematical lexicon allow us to engage with criticality in a manner that does not amount to judgement immediately— always only situationally, temporarily, and contractually so. Rather than providing for "good" judgement, criticality provides for experience. On the basis of experience, mathematics in Serres's philosophical language (i.e., one of "many portals") allows for treating the two (reason and justice) as parallel structures that will never exhaustively coextend. This is the argument that Serres developed in *The Natural Contract* (1990). We will only touch upon it briefly here.[35] As parallel structures, reason and justice actively relate to each other: they engender each other in their ever-evolving but entirely objective faculties—*faculties* in the literal sense of what supports a capability of bonding, bounding, relating objects (i.e., not as the faculties of a generic [human] subject but as the capacities and capabilities [in precision, rigor, and finesse] of an objectively formulated natural contract). We could call this an engendering of *facultativity*. We can think about *reason* and *reasonability* through criticality, that is, about how criticality relates to *invention:* invention is reason rendered facultative. Criticality, in this setup, is the mathematical inverse to a domain of judgement. As such, criticality (couched in terms of what is trustworthy rather than true or false, probable or unlikely) can form a parallel structure that lives from, and feeds into, a manner of thinking that is inventive.[36] Rather than deciding *whether* things are to count as compatible (if they work objectively, if they *function* or not) mathematical realism establishes *that* and *how* things *can coexist* in an decoupled but reasonable and compatible manner. Such compatibility is something that must be achieved, acquired, and maintained *objectively.* Treating two structures in parallel means to *instruct a third* that provides a more abstract domain capable of accommodating both.[37] This third structure needs again to be submitted to the critical-inventive interplay, and thus the third never ceases to be further instructed. This *third* domain is what we could call the impersonal and active referent for a generic notion of culture, the natural referent of the social. Such compatibility is real, and objective. It is not what depends upon an individual's reasoning to be maintained subjectively. In the most radical manner, this novel sense of intuition that Serres was enthusiastic about, and which his oeuvre sought to help emerge and prosper, was a sense of intuition that is *proper to things themselves*, it was a sense proper to objects coexisting among objects. Such objectivity is not "pure." Nature articulates itself, to Serres, as "a contract between the Earth and its inhabitants."[38] The earth as a planet, a terrestrial body, is without doubt universal. But it does not represent the universe; rather, it embodies it in a particular way—as do all natural things. Such a reconciliation between truth and sense requires that we pose the question of truth, not in relation to appearance and falsity, but by relating it, through form, to death. Hence, Serres thought of his "epistemology" also as a general treatise on sculpture. For Serres, a philosopher is someone who is capable of *anticipating* the world of tomorrow.[39] A philosopher is not a judge, nor an advocate or a prosecutor. Philosophy, to Serres, is midwifery. He continued the dialectic legacy of philosophy as dialogue; but since form cannot be purified by deciding about what is noise with regard to it, noise is to be kept contained (and capable of being remembered)—*buried*. The objective domain that is to grant decision needs to be dispositioned itself *formally*, in mathematical terms inverse to the domain it is to grant decidability. Hence in this form of dialectic realism, the two interlocutors are not on opposite sides. Rather, they are on the same side: the iconic image through which Serres illustrated this was Goya's painting where two men are fighting each other, possibly with a crowd of people watching eagerly to see who is winning—without realizing that they are standing in quicksand.[40]

For the mathematical realist, then, who is at the same time a mathematical mystic, reason must be addressed by the impersonal pronoun "it"—the voice that reason is to respond to is that capable of articulating a relation of equipollence between the real and the rational. I will describe the "subject" behind this voice by a particular model of communication which Serres introduced in *The Parasite*, and which he calls "that of the Paraclete," literally "the one who can be called to one's aid" (Greek *parakletos*, for "advocate, intercessor, legal assistant," and from *parakalein*, "to call to one's aid"). We have in Serres's philosophy a notion of nature that is the object of science; hence anyone or anything must *in principle* be welcome to step into the central position within this domain of reason's responsibility. Nature is the voice that speaks in all tongues, and that never ceases to speak originally. Nature speaks physically, in the terms of collective forms of behavior that regard the "triple tress" of "information, energy, and form" that bond "knowledge, power, and complexity" in *The Natural Contract*.[41] Serres's is a philosophy of nature, and his nature is contractual in the sense of a communicational physics. The core *political* idea of what Serres called "the natural contract" is that we need to think about reason in terms of principality as much as we need to think about principles in terms of reason.[42] When we address reason as a principle, when we personify it *inversively*, that is as the real referent for the impersonal pronoun "it," then we can couple it with judgement. Then reason can produce cases, but the cases, even though they are more or less well reasoned, will be arbitrary, "Hierarchy remains inside reason, but since height, power or king are no longer spoken of, it becomes transparent inside reason, so invisible that no one has seen it."[43] If we want reason *without an arbiter*, without a judge, and if reason is to rationalize something mystical, then something mystical must be attributed to this reason in turn (its rationality must count, to a certain extent, as mystical). It is in this sense that, as Serres puts it, "The principle of reason demands that there be no reason."[44] This principle of reason is the principality whose rule is acceptable to the realist mystic of mathematics because it demands respect, but does not ask for being acknowledged as legitimate domination. "It is better to make peace between the two types of reason in conflict today" (judgement and critical reason, standing for tradition and innovation respectively) "because their fates are henceforth crossed and blended and because our own fate depends upon their alliance. Through a new call to globality, we need to invent a reason that is both rational and steady, one that thinks truthfully while judging prudently."[45]

Since Georg W. F. Hegel, dialectics has been usually thought of as a spiral process. Serres discretized such spirality through the operative role he attributed to code: Nature is universal and principled, he affirmed, but it speaks in the heterogeneous scales of all that can be *sounded* by inventively-critical reason. This requires thinking the inverse of the spiral form: thinking of the inverse in this manner elaborated earlier is a spiral process as well. But instead of *revolution* (redistribution of head and base positions) we have a kind of *inverting*, of turning inside out (what is called *eine Stülpung* in German). This is what Serres considers "a new birth of a new world." It is reasonable only in the sculptural form of silent words that behave like statues. We can think of such nature as a generalized sense of rebirth.[46] The hierarchy between logics and mathematics, between reference and projection, can never be considered settled once and for all. In this sense, the sole principle of reason Serres was ready to affirm—and this indeed is the key motif in his foundational triad[47]—was what takes as its reference (think: logics) an indifference (think: mathematics) toward all difference (think: experience, real phenomena). "Is reason defined by indifference toward all difference?"[48] he asks. And he puts it even more drastically: "Reason demands that there be no reason."[49] This is how a natural philosophy can claim to be *universal* (pertaining to all things) if the universe is what "holds by this principle without principality."[50] This indeed is why it is important to see how Serres reconnected with the ancient tradition of meteorologica, and their concept of "commensurable universals." When Serres proclaims that there is an intelligence immanent to and coextensive with the universe, he is actually suggesting how to address (metaphysically, categorically) this locus where the place of this principle is. This place is by necessity a vicarious place, a place that can only ever be occupied by substitutes, by placeholders. We will see how this means that we ought to relate truth to death rather than to falsity, and that we ought to consider death as the "origin" of form. This is what Serres's Logos of episteme, its formulation as a treatise of sculpture-in-general, is all about. The kind of epistemic optimism that Serres embodied, perhaps like no one else in our current times, draws from this. To fathom and demonstrate what such optimism entails is what this book seeks to achieve.

The Plan of This Book

Chapter 2, "Quantum Literacy," is organized under two sections: "Elementary Indecision" and "Taking Ignorance into Account: Quantifying Strangeness." It introduces a group of rather technical concepts (ciphers, entropy, negentropy, the Price of Information, invariance and teleonomy, and objective chance) that are crucial for appreciating the framing idea around which this book is organized—that any philosophy ought to always maintain compatibility with science, not serve or seek to dominate it. In order to be able to achieve this today, philosophy ought to change the paradigm from materialism in terms of "production" to one in terms of "communication." Perhaps the most important shift this entails is in learning to cope with a certain notion of indeterminateness and cyclicity that is immanent to materiality itself, and within which chance factors as an objective, formally determinative aspect. This chapter will address how with information theory, we can think of mathematics as a language that provides a lexicon whose words articulate silence against the background noise of the rumbling mass. Those words can be "spoken" (indeed: "pronounced" through code) like musical instruments can be played, but only in an inverse fashion: we can learn how to play them as instruments that articulate silence (rather than sound). Such "speech" is what is introduced here as quantum literacy. How speech in such literacy articulates language as *puissance* will be discussed. At stake is a language that "can," not one that "is" ("*La langue n'est pas, elle peut*").[51] Such an understanding of language affords a novel theory not only of objects, but also of subjects. According to the inversive gesture that such a change in paradigm invites us to embrace, objects can be quantified in their rarity and strangeness rather than in their regularity. With regard to the subject position, it will be argued that quantum literacy is entirely about "speaking objectively among objects." Like the pronunciation of silence, the occupation of this subject position is one of undoing belongingnesses through quantifying them. Subjectivity in such speech is engendered through an active *not-identifying* with one's own articulations. Such a passively active stance of speaking-without-identifying individually with the uttered words will be tentatively characterized through a particular form of bearing witness, because there is a singularity to this impersonal subject position. Through the ethical commitment of this literacy to speaking objectively, this subjectivity ought to count as universal and yet, also as entirely situational. At the center of power in terms of this new materialism, the master and subject mix and coincide.

Chapter 3, "Chronopedia I: Counting Time," is structured in two parts: "Meteora: The Wisdom of the Weather" and "A Logos Genuine to the World: 'Le logiciél intra-matériel'," a logicial formality that is immanent to matter. It discusses how a theory of the subject position according to the quantum literacy (as proposed in Chapter 2) would be futile were it not to come with a novel conception of history. Such a novel conception of history is proposed with respect to the fact that since very recently, every discipline of the sciences has started to *date* their objects, to place them within a universal history (that which recounts the age of the universe). Against this background, this chapter introduces Serres's important notion of the *Chronopedia*. Its core concern is a kind of general education that is public and common, much like it was the preoccupation of Varro, Cicero, and Vitruv in Roman rhetoric when they organized education according to an *encyclopedic* notion of knowledge. *Encyclopedia* literally means "a general course of instruction," from Greek *enkyklios paideia*, taken by Latin authors as "general education," but literally meaning "training in a circle," that is, the "circle" of arts and sciences, from *enkyklios* "circular," also "general" (from *en* "in" and *kyklos* "circle"). We will discuss how we cannot speak today about general education through the figure of the circle, because it cannot capture "ageing," the "massive" passing of time, time as passing in a fashion that is of material consequence. It will be argued that today's science relates to the universe in its aspect of *ageing*, and that accordingly, the education of universal knowledge too must be capable of accounting for this. To this end, the notion of the *Chronopedia* suggests that we need to place our knowledge not in a circle directly, but mediately so, by packaging it in discretized and countable time (*chronos*). This chapter hence introduces the key notions for counting and the measurement of time in relation to world models that connect "weathering" with the "counting of time": The central notions therefore are those of "scalarity" as a scaling based on modules. Such a notion of scale is disconnected from any sense of global dimensionality. Like this, it can accommodate its own inversion. Other key notions will be those of "chronology," "meteorology," and the sundial (the "gnomon").

Chapter 4, "Chronopedia II: Treasuring Time," is somewhat of an exercise in concentration. Its subsections are "Homothesis as the Locus in Quo of the Universal's Presence" and "The Amorous Nature of Intellectual Conception: Silent Words That Conserve the Articulations of an Impersonal Voice." It is a reading of Serres's early article on the legends of Thales, and about what he imagines that Thales saw at the foot of the Egyptian pyramid when he learned to stop time in order to inflate one (timeless) moment. We will discuss how Thales thereby obtained a space for abstract "likeness" (the space of *homeothesis*), where two unlike things can be treated within one common projective scope, whereby neither one of those two things needs to be reduced to and subjected to the identity-domain of the other. This chapter demonstrates how, at the heart of geometry as we can learn to think of it with Serres's philosophy, there is a wealth to be "sourced" that can never exhaustively be treasured. Such wealth is intellectual as well as material, and it can be tapped into by paying respect to what one neither understands nor likes. It will be demonstrated how the lexicon that unfolds from mathematical language is partly resulting from technology. With its concepts, the theater of measurement closes its doors, Serres maintains, and the world begins to cast shadows once again.

Chapter 5, "Banking Universality: The Magnitudes of Ageing," is divided into two parts, "Metaphysics" and "Freedom." Philosophy should ask along with the sciences: "How old is the world?" In this fashion, Serres speaks of a "Real Age" of the world that is common to "old men and the newly borne, grandchildren and grandmothers, animals and plants, friends and enemies, to all that are carrying DNA."[52] He thereby gives us a model of how: (a) our thinking draws depth from being embodied; and (b) how to hold on to the idea of generational descent in terms of *dia-sequentiality* (i.e., without submitting to the linear branching of the tree as a law of *con-sequentiality*). With respect to time, all beings at any instant are to be considered as equal—an equality that is contracted "in two fractions: one minimal, the individual age, the other much larger," Serres speaks of it as "the universal belonging" to "La Grande Vieillesse." We will discuss how Serres introduces thereby a way of complementing Darwinian evolution of speciation and differentiation with an inverse one of de-speciation and de-differentiation. This latter is termed Exo-Darwinism, and it is crucial for Serres's rejection of seeing the natural and the artificial as either in a categorical or a dialectical opposition.

Chapter 6, "The Incandescent Paraclete: Tables of Plenty," is divided into two sections: "Equatoriality Generalized" and "Generational Con-sequentiality." It picks up the line of dia-sequentiality with regard to the family model of descent, and relates it to a particular geophilosophical imaginary. The chapter discusses Serres's genuinely abstract approach to the problem of foundations of knowledge in his trilogy *Rome* (1983), *Statues* (1987), and *Geometry* (1993), and demonstrates how he raises the problem of "foundations" to the scope of a *generalized equatoriality* that entails interesting proposals for viewing the world beyond any notion of the "soil of the ancestors." The Orient and Occident feature thereby in an a-territorial geography that commits itself to a departure and break from any form of accounting to chthonic soil—whether autochthonic or otherwise. Such an a-territorial geographical philosophy considers "Occidentality" and "Orientality" as two poles that distribute, through a generalized notion of "horizontality" and a pluralist notion of quantifiable "verticality" (the "scalarities" of the meteora), everywhere throughout the world. We will discuss how Serres picks up the tradition of the ancient meteorologists who spoke of a *mixed human condition*. According to a wisdom of the meteora, earthly conditions (geographical, geological, practical/technical) shape a living climate together with cosmic (astronomical, astrological, theological) conditions. Serres's interest with such an a-territorial form of geographical ideation, it will be argued, is to reconnect with this tradition from a contemporary Anthropocene perspective. This idea will be discussed by relating a particular model of communication as a system, which Serres introduces in *The Parasite*, that of "The Invention of the Paraclete, on the Pentecost," with another model Serres introduces in *Hermes I*, entitled "Le réseau de communication: Pénélope." The chapter discusses how a philosophical discourse committed to such a-territorial geographical thought ought to be one that is not at home in the reason by which it lets itself be guided. It will be introduced how Serres proposes to speak of such discourse as "exodic," in contradistinction to "methodic."

Chapter 7, "Sophistication and Anamnesis: Remembering an Abundant Past," is divided into two sections: "Architectonics within the Domain of a Withholding Power" and "Classicism: Remembering Contemporaneity." It discusses the theorem of the Price of Information in relation to a study by Marcel Hénaff on philosophy's relation to money and the problem of the convertibility of truth. It begins by foregrounding the question of the priceless, and asks: If knowledge is not to be so radically quantified (as a quantum literacy proposes), does this not mean that we inevitably reduce our relation to the world and to each other to the quantifiable management of goods, up to the point that the spiritual icon of the priceless—a safeguard against hubris and self-righteousness—is finally annihilated, leaving a cold objective world bare of anything we might still call humaneness, modesty, grace and pity? Against this background, the chapter presents a close reading of Serres's article "Structure et importation: des mathématiques aux mythes" (1968). In it, Serres proposes that the algebraic notion of "structure" allows us to relate to myth via mathematics. We discuss Serres's relation to structuralism and post-structuralism, especially to his teachers in the history of ideas, Georges Canguilhelm and Gaston Bachelard. Serres shares with them the interest of finding a new relation to classicism in order to hold on to the notion of the priceless. The distinctive aspect in Serres's interest in structure is that he regards it as something like the formal *syntax of metaphoricity,* of which he claims that it is always already at work in myth. As is claimed for metaphors in poetic language, Serres also claims for structures in mathematical language that they facilitate "transport" that involves some kind of import/export of contents; in the case of mathematical language we propose to call this import/export "metaphysical" because it involves most abstract conceptions like those of finitude and infinity, boundedness and openness, and discreteness and continuity. At stake in this interest in the relation between classicism and pricelessness is a form of "existential trust," which will be discussed by looking at Roland Barthes's iconic text on the Eiffel Tower. We will contrast Barthes's approach to "really building a cipher" (as a *modern myth*) to Serres's approach to "really build models that communicate" (as a *realist classicism*).

Chapter 8, "Coda: Architecture in the Meteora." Serres's view on the coefficiency of math and myth in his "realist classicism" entails that there is universality to culture. On the basis of this, Serres is ready to grant that all things may be miracles; but this does not mean they cannot be treated reasonably, or objectively. The last chapter ponders on how what is at stake with this claim is a particular form of optimism we suggest to call "epistemic optimism." It goes along with an architectonic of an abundant reason, complemented by quantum literacy. "If it is not a miracle, then can we build it?" Serres asks repeatedly. We will discuss how with respect to such a notion of "building," an eminently political role of architecture might once again begin to play against the "dictates" of economy. Surprisingly, *copia* is the Latin term for "plenty," and *copiare* literally means "to transcribe," "writing in plenty." It also means "duplicating," but through this emphasis on the plenty, any relation between "originality and duplication" is no longer one of "closest possible representation," it is no longer guided by the ideal of rendering the difference between copy and original transparent. Architecture in the meteora, hence, needs to be masterfully "copious." It projects the domain of the quasi as a real and feasible locus in quo for the coexistence of as many forms of "commensurate universals" as possible. In other words, it is committed to "serving nature through mastering it," in order to coexist as nature with nature. At stake is a political mastery that governs nature by striving to let it "be," to not have to touch and exploit it, because it is possible to "copy" it. What such mastership is committed to is, in short, to a novel form of politics, not economy.

Chapter 0, "Instead of a Conclusion: The Static Tripod." The reason why the book finishes with a "cipher" chapter (number zero) is to reinforce the overall gesture of Serres's approach to time, namely, to engage with an abundant past in order to keep futures open. It is a long excerpt from Serres's second book on foundations, *Statues,* entitled "The Static Tripod." The excerpt is a dramatic dialogue between Oedipus and the Sphinx, on the unsettling but not inimical implications of Serres's thinking for a non-anthropocentric and inchoative understanding of human nature in the terms of hominescence.

2

Quantum Literacy

I'm in the habit of trusting Hermes. He is the god of codes and secrets.

Michel Serres, *Rome*[1]

This chapter elaborates Serres's postulate that we can cast off a world of myth, as well as one of pure reason, if we learn to be literate in the language of mathematics. Serres's core interest was to respond to the unsettling widening gap between science and culture by finding a novel basis for criticality, and a novel grasp of a sense of intuition; one that would not be at odds with the counterintuitiveness that accompanies much of twentieth-century science and the technicality of the objects that we all handle daily. It must be an embodied sense of intuition, and yet one with which we are not always already equipped. The sense of intuition at stake is one that must be acquired. Hence, I want to frame it here in the terms of a novel kind of "literacy" in the language of mathematics. But it cannot just be a literacy about objects. It is also a visceral kind of literacy, one that that affects one's guts, ones' innermost, physical and embodied sense of proprioception. My proposal is to frame such literacy through relating what I call "a form of bearing witness" with "sophistication." There is one particular quote from Gilbert Simondon that I would like to use as a motivic key here (even though this book is not the place to discuss Simondon's important oeuvre) because it captures well the perhaps most challenging aspects we will be concerned with in this chapter. It is from an interview with Anita Kechickian entitled "Saving the Technical Object": [2]

> Anita Kechickian: In 1958 you wrote about alienation produced by non-knowledge of the technical object. Do you always have this in mind as you continue your research?

> Gilbert Simondon: Yes, but I amplify it by saying that the technical object must be saved. It must be rescued from its current status, which is miserable and unjust. This status of alienation lies, in part, with notable authors such as Ducrocq, who speaks of "technical slaves." It is necessary to change the conditions in which it is located, in which it is produced and where it is used primarily because it is used in a degrading manner. … It's a question of saving the technical object, just as it is the question of human salvation in the Scriptures.[3]

With our interest here in quantum literacy we will bracket Simondon's question of salvation with regard to the technical object by attending to the two roles of bearing witness and literacy that are, arguably, at work therein.

Elementary Indecision

Bearing Witness and Literacy

"Bearing witness" is a difficult term to use in philosophy, as indeed is "literacy." Jean-François Lyotard, with whom Serres shared a broad impatience over the lack of attention paid by intellectuals to communication in its own terms, wrote an essay originally entitled "Le non sens commun," which he ends with the sentence "the witness is a traitor."[4] In this article, Lyotard's theme is the same as Serres's: Can we find and develop a kind of intuition, a transcendental criticality, that could support a *commonality* among people because it is derivative to a common time, a common space, a common sense (Lyotard's term is "*un lieu commun*"). For Lyotard, the basis for such commonality would be to know suffering *affectively*, as something that *exists* impersonally, as something whose cause ultimately escapes an individual's faculties of reason. Hence, with respect to the individual, it is a transcendent sense of an impersonal suffering. Lyotard gives credit thereby to something of which he holds that it stems from an inappropriable source and is hence something priceless, and ultimately not tameable. The dilemma he addresses is that to reason intellectually about real, existent suffering appears to eradicate, to annihilate, this very basis for humaneness. Science, and the increasing level of comfort its knowledge provides for living together de facto numbs the affective basis for "knowing suffering" impersonally. For Lyotard, bearing witness inevitably contributes to the annihilation of that which cannot be domesticated, tamed. His closing sentence, that the witness is a traitor, seems categorically to reject the path chosen by Serres. For Serres, it is possible to actively reason while *remaining undecided*, as a way of bearing witness without betraying. It is a way of bearing witness *objectively*, as we will see. For him too, it is something inappropriable, visceral, and affective that can serve as a support for commonality. Much like Lyotard's interest in an impersonal experience-based knowing of suffering, the sense of intuition Serres's philosophy promotes is not one that pertains to an individual subject. For Serres, as for Lyotard, this sense of intuition had to be impersonal. But for Serres, relating the bearing of witness to literacy must not only be impersonal but also *anonymous*. What is at stake with the kind of literacy in the language of mathematics is a literacy that would not pertain to an individual subjects' faculties. It is a literacy in which "one" speaks "anonymously," neutrally, as "no one in particular" because it could be "anyone" who says it; and yet there is contingency to what can be said this in language. We will discuss shortly how this latter aspect, anonymity, sheds a different light on the activity of remaining undecided. Serres's sense of intuition is *objective* in the sense that it does not presume, or originate within, an individual cogito, equipped with a set of faculties of limited (human) capacities. The subject of Serres's objective sense of intuition—the agency that responds affectively, as well as intellectually (with literacy), to this sense is neither an active agency ("I") nor a passive owner of this sense ("me"). It is the anonymous third-person singular, the "it" of an impersonal cogito: the more we are thinking, the less we are "us," Serres maintains. The more we are thinking, the more anonymous we become.[5] Anonymity is something that must be achieved, acquired.

But before going into this, let us look at it from another angle. Is a sense of intuition not about providing a source of security, stability, or an uncorrupted reference? Clearly yes, but then we need to ask: Would such a sense, with which the more strongly we connect the more we cease to be us, not yield quite contrary effects? How could knowledge, supported by a sense of intuition that consists in actively acquiring not only impersonality but also anonymity, possibly strengthen a sense of responsibility with respect to its own status? The argument is complex and involves meditations on the notions of identity, totality, and order, but also on authority and authorship, as well as on the sacred and the spiritual. Serres offered some insight into his views on these last two ideas in a recent interview. He addressed the proposition that science alone is incapable of fulfilling the societal role previously occupied by religion, leaving a society stripped of spirituality and largely without orientation and direction. Serres's response is especially interesting in that he decoupled spirituality from the sacred, by relating both to the problem of violence: "René Girard has given us a sense of the sacred which is very precise: sacred means sacrifice." Our civilizations are far from having overcome practices of sacrifice, he maintains—every civilization struggles with the problem of violence, and this entails the problem of violence's sacrality. Serres calls sacrifice the "archaism of civilizations":[6] Every form of violence is a readiness to sacrifice, in the name of a particular rationality. His crediting of the impersonal third, the anonymous and impersonal cogito of the "it," is related to a certain understanding of spirituality as *the active cultivation of indecision*. The cultivation of indecision is the point of entry for responsibility. The objective sense of intuition, which is to guide our practices of knowing as those practices where none of us is an "I" or a "me," but an anonymous "it," draws attention to the backgrounds of *non-knowledge* that accompany any form of knowledge whatsoever. There is a spirituality to this objective sense of intuition that manifests itself in the primary practice of bearing witness in the language of mathematics as something that requires literacy. The most important aspect of it is that remaining undecided is not a comfortable stance, but involves an active withholding and is, hence, difficult and exhausting to achieve. Literacy, here, is to be understood in relation to a "natural" puissance, and as such, it is a power but not one that could be sure of itself. Literacy is in service to the world, and it is related to a natural puissance of undirected (because it is omnidirectional) activity. Bearing witness cannot be reduced to a skill and its technical mastery in terms of either/or. It can be acquired and interiorized in a myriad of ways—just like learning to write or playing a musical instrument can, and at a similar personal "cost" (namely, the submission to exposure, practice, and discipline).[7] Bearing witness, if it is not to be the act of a traitor, is a matter of abstraction and technicality as much as of finesse and sensibility. Such a philosophy's relationship to truth is neither one of prophecy nor one of revelation. It is a philosophy of *demonstration*—but one where demonstration and proof are architectonic. Demonstration does not expose truth nakedly. Every demonstration is at once a revealing and a veiling articulation of something meaningful that is being witnessed—meaningful in the sense of *important*. Literacy manifests in knowing how to grasp intuitively, affectively, things in a scope that is situationally adequate and appropriate. It manifests in having a "good" and a "common" sense for how to rationalize the scope of their *importance*, literally by grasping how to *import* contents that factor in one's intuitive grasp from one "locus/locution" (situation of insight and experience) into another.[8] As such, literacy manifests itself as the inverse (not the opposite) of explanation: when, as we propose here, the intellect's concern in bearing witness is an *import*, we can think of the intellect's concern when giving an explanation as an *export*. Import and export of what? Contingency and arbitrariness, manners of discretion. When explaining, we seek to purify one rationale (one model) of arbitrariness. When bearing witness, one seeks to discredit one's own inclinations that inevitably always characterize explanations (those which appear acceptable to one's sense of righteousness, whether immediate or reasoned). Such a notion of bearing witness involves a notion of neutrality that is not virginal and pure, but one that is, so to speak, of universal genitality because it interiorizes virtually any specific nature's sex.[9]

Actively affirming a stance of indecision does not amount to passivity, because it requires a critical form of reasoning that is not the rational (de/finite) other to spirituality (being the infinitary complement to the rational), but one in which reasoning is spiritual and in which the spiritual is decoupled from the sacred. It requires a form of reasoning that proceeds rationally but without annihilation, sacrifice, or purification. For Lyotard as for Serres, such a form of reasoning is algebraic and economic (Lyotard's *Libidinal Economy* (1974), Serres's *The Parasite* (1980)); both accordingly attend to the House, the Oikos, the Domus, and Domesticity, to place the idea. But Serres thinks about the architectonic of such a domesticity via mathematics, and thereby positions the anonymous, impersonal cogito that must be acquired and achieved, as the master of mathematical reason, within an economy of maxima and minima, more or less, rather than either/or. The *lieu commun* (the common space and time) established by mathematics is not a transcendent reference for real knowledge; rather, it is the generous transcendental domain, where reason indexes singular realities by cultivating indecision, by seeking to formulate demonstrably what it witnesses, objectively. In this way the sense of intuition at stake here does not eradicate a sense of responsibility but strengthens it—and this even though the more a master "thinks," the less she or he is herself or himself. This sense of responsibility then relates to a notion of authority in the double sense of the word which Serres elucidates when he mobilizes a set of interrelated meanings around the morpheme "au":[10] that of a legal guarantor of something, like the *author* who takes responsibility for what she writes by signing it, as well as the meaning which comes from *augmentation*, and *auction*, and which means to raise the worth of something by giving credit to it, and thereby augmenting it. A teacher, like a parent, gains authority by "augmenting" the students in relating to what they themselves already have the disposition and capacity for, especially if they do so without realizing or having confidence in that they do. The book where Serres develops this idea the most is his book on learning and education, *Le Tiers-Instruit* (1991). But let us look closer at how to imagine such "cultivation of indecision" and its architectonic.

Cultivating Indecision: The Quantum Domain's Domesticity

For philosophy to maintain compatibility with contemporary science means maintaining compatibility with quantum physics. Yet what sense can we make of this quest for an objective sense of intuition that is proper to the things of the world, and for something like "quantum literacy"? We know, disturbingly well, that what makes the quantum domain so difficult and indeed counterintuitive for our epistemological habits is precisely its odd condition; at a very elementary or ground level, it is characterized by indecision. To promote a sense of intuition with quantum physics sounds like a logical absurdity. In quantum physics, there is a principal and objective indecision with regard to whether one is dealing with matter or light, particles or waves, figure or element, location or distribution. For all these couples, this indecision obliges us to take both poles into account simultaneously. It is not possible to subordinate one to the other in any one particular hierarchical way; explanations related but not identical to the point I wish to make here by developing Serres's notion of neutrality are speaking of "entanglement" or "exchange of particle charges at subatomic scales." The important aspect for us here is that even though with our inherited epistemological categories these couples must count as *opposites*, they ought also to be regarded as poles of cycles, that is as implicatives of "commensurable universals" that act reciprocally upon each other by thwarting discretely, while at the same time complicating continuously, each other's courses—like the ancient meteorology tradition was used to think, or like mirroring images do within a broken symmetry no longer pure but marked by passing time (we will discuss this when we attend to Serres's communicational physics).[11] This way of thinking about opposites as actively reciprocating poles will be discussed in Chapter 6 as the "generalized equatoriality" as the "incandescent body" of the "invented Paraclete."

What I call in the title of this section "the quantum domain's domesticity" is not only to stress that nature, especially from the point of view of quantum physics, behaves economically (and hence rationally, reasonably). It is also to illustrate that quantum literacy involves a kind of thought in which the thinking subject cannot possibly expect to feel "at home." One of quantum physics' greatest scientists, Richard Feynman, used to begin his lectures with a warning: he would tell his audience to beware, that his subject matter could not possibly be comprehensible, that he did not even understand it himself.[12] Yet, Feynman would go on, his account should by all objective means be considered as relying on rigorously appropriate depictions for a certain kind of strange experience that can be relied on, and used in calculations. They display a precision, after all, that allows us to design devices with controllable dispositions at the quantum level such that: electricity can be garnered from the sun (photovoltaics); or makes it possible to set up abstract molecular formations for the cloning of living organs and organisms (molecular biology); or leads to the production of semiconductors in which the circulation of electrical current can be controlled in all the manners of coding that propel information and communication technologies. It is worth looking more closely at how Feynman used to lecture on "what nobody can understand anyway." The topic of the lecture was "The Character of Physical Law," and Feynman told his audience that this lecture "is perhaps the most difficult lecture of the series, in the sense that it is abstract, in the sense that it is not close to experience."[13] If he were to talk about his topic without any description of the actual behavior of particles on a small scale, he explained, he would "certainly not be doing the job."[14] This is interesting because it illustrates very well what kind of witnessing was at stake: Feynman professes, and somewhat subjects his audience to, a certain discomfort. As he saw it, there is no way around it: "This thing is completely characteristic of all of the particles of nature, and of a universal character, so if you want to hear about the character of physical law it is essential to talk about this particular aspect."[15]

Feynman continued: "It will be difficult. But the difficulty really is psychological and exists in the perpetual torment that results from your saying to yourself, 'But how can it be like that?' which is a reflection of uncontrolled but utterly vain desire to see it in terms of something familiar."[16] This empathy with the audience's likely confusion only extends so far. He continued: "I will not describe it in terms of an analogy with something familiar; I will simply describe it."[17] He will not revert to a familiar model first, so the reader could better know what to expect. He was "simply going to tell ... what nature behaves like,"[18]—he would do this "by a mixture of analogy and contrast,"[19] rather than by offering a theory. Crucially this means that there would be gaps in how to make sense. Rather than aiming at an explanation, or by providing a descriptive model, Feynman's manner of description remains puzzling, and finds its voice, its articulation, by giving versions and working out contrasts between them as a means of helping the nonexpert audience grasp as much as possible of nature's "delightful, entrancing character,"[20] to which he bore witness in his work. We find in another book by Feynman a helpful explanation of this tactic, one that makes clear that this is *exactly not* what it seems; it is *not* a form of irresponsible entertainment, as some might see in Feynman's joviality, but the opposite—it is the active *frustration* of the audience's desire to be comfortably entertained:

> What I'd like to talk about is a part of physics that is known, rather than a part that is unknown. People are always asking for the latest developments in the unification of this theory with that theory, and they don't give us a chance to tell them anything about one of the theories that we know pretty well. They always want to know things that we don't know. So, rather than confound you with a lot of half-cooked, partially analyzed theories, I would like to tell you about a subject that has been very thoroughly analyzed.[21]

Feynman's advice to his audience simply was, "not [to] take the lecture too seriously, feeling that you really have to understand in terms of some model what I am going to describe, but just relax and enjoy it."[22] He was going to tell them what nature behaves like, and "if you will simply admit that maybe she does behave like this, you will find her a delightful, entrancing thing."[23] We will not look more closely here at how he described this behavior (although it would be interesting). Instead, we will introduce some basic distinctions constitutive for what is proposed here as a novel literacy, a *quantum literacy*. The relations between bearing witness and literacy offer us a different grasp on mechanics as an art of formulating *states of affairs* objectively, which is the underlying interest in this chapter.

Ciphers, Zeroness, Equations: Architectonics of Nothing

On the elementary level of quantum physics, it seems that we must acknowledge, that there is a kind of indecision that is principle, physical, objective, and real. It is not merely a subjective effect of insufficiently grasping the complexity of phenomena in our epistemological frameworks. Although both, the failure of grasping complexity and inadequate epistemology are likely to exist, there remains a physical and objective dimension that persists: it is by operating rigorously (mathematically) with the non-knowledge of objective and principle *indetermination* that the repetition of effects can be triggered, and the reproduction of constellations can be achieved in technical descriptions as well as in technical objects. I use the term "constellation" here deliberately to comprehend both physical and ideational content, the point being that matter and code can no longer be purified against each other, as in looking for *pure* code and *pure* matter. While the view from classical physics would easily align matter with particles and light with waves, in quantum physics the alignment is not so easy: quantum electrodynamics indeed regards light as particles, while matter is regarded as waves (in terms of phases and frequencies).[24] But even a contrast like this is misleading, since perhaps the most disturbing point about the quantum domain in physics is exactly that light can no longer be thought of as the immaterial complement to matter: the sun is, quite literally, a massive and brilliant ball that emits, in the white light of its radiation, quanta of electrical charges—white matter! These quanta come assorted and packaged as atoms, molecules and all that comprises them. The "elementary" level in the quantum physical domain, that of the quanta of electrical charges, must count as indetermined, as only spectrally apparent mixtures that come in all hues, colors, and frequencies. This level can virtually be traced in any possible thing at all. It is the zeroness, the origin, of all that will ever have mattered (within one solar system, at least) in the same sense as white light is the zeroness of colors: not because white light is bereft of color, but because it is so abundantly full of color. White light is white because it comprehends any color that might come to the fore, once some of the radiating quanta are absorbed, through whatever circumstance. In the quantum domain the elementary level is, so to speak, where only jokers are playing, where only the substitutional operator is at work, the agency of exchange, that of the algebraic symbol. For Serres, algebra is where tautology coextends with the mythic, where originality coextends with history[25] and where the relation between the regions of locality as well as versions of a global integrality all find accommodation within a universal domesticity provided by natural communication.[26]

Serres's philosophy is the architectonic of a domesticity that is universal, as we have suggested. It is an architectonics relative to what we called earlier in this chapter "the quantum domain's domesticity," as the domesticity of a domus in which we can never hope to feel at home with our personal, individual thinking, but to which we, like everything else beings of universal nature, nevertheless "belong." Serres philosophy is the architectonic of a domesticity that it universal, hence, but only *legendarily* so—and legends, he maintains, "are always written in a metalanguage." They "intervene in the middle of a legendary narrative with a change of language, a change of register, tone, a change of scale, state, space and time."[27] The content of legends is always "hidden, coded, to be deciphered":[28] "Yes, legend indeed means: how to read what is to be read."[29] The quantum domain is, architectonically, the domesticity of travelers, messengers, of beings in transit.[30] The figure of Hermes, and the language of medieval angelology, lend themselves well to talk about communication, Serres maintains, because it was concerned with angelic beings that *do* have sex, that *can* fall in love, wherever they pass through, even though their being is spiritual in kind. He finds here a language to speak, legendarily, in terms of a kind of lovemaking where sex is decoupled from reproduction—lovemaking in the age of the pill, Serres maintains.[31] Natural communication, as the domesticity of the universe's sex, is the house and home of a kind of filiation that originates not in a tribe or in a race, neither in territorial nor blood roots. It is a domesticity that is genuinely *transcendental* in kind, a domesticity where the conditions of possibility for the experience of things model those things— but these conditions of possibility do not pertain to a subject, like in Immanuel Kant's or Edmund Husserl's transcendentalism: "When reflexive epistemology becomes intrinsic, the field of the transcendental passes over to the objective."[32] The domesticity of the universe's sex is a domesticity that manifests in objects among objects. It is a domesticity that *engenders* the subjects it welcomes and accommodates. It is (we might say in resonance with Baruch de Spinoza) a House of Joy: the transcendental locus and home of a passion through which the spirit passes to greater perfection.[33] In Serres's transcendental philosophy Hermes, messenger and wanderer, god of the crossroads and merchants, is no longer the antagonist of Hestia, goddess of the foyer, the home and the hearth. Rather, Hermes and Hestia are lovers; together they are capable of an epistemology where *episteme* does not so much mean "knowledge" but, formally and methodically, a domain of indefinite *invariance* and *stability*. In this epistemology that provides for universal domesticity, "science gives itself invariance as a rule."[34] The conditions of possibility for experience, "occur in the world. It is no longer we who inform the world with our schemas and our subjective categories, the world itself is information."[35] It is the epistemology of a transcendental philosophy that is not rooted in a notion of the subject. Between Edmund Husserl and Léon Brillouin, according to Serres, we have learned a lot about the conditions of possibility of *things*: "The new science of objects obliges us to make this philosophy here,"[36] but we need to be precise: "things are not only information." Take flowers, for example, he suggests. They carry within themselves a "hard" constitution, one with energy, one which enables them, for example, to capture some light from the sun and transform it into energy. Flowers, like all things, are "energetic machines." They are energetic, but they are also information. And this information is decisive for the *transfer* of energy: "It is precisely because chlorophyll has this or that chemical formula, which is of the order of information, that it can transform light to matter."[37] We can easily recognize here Feynman's concern with the character of the laws of physics—when he remarks, somewhat mischievously, that he can safely say that nobody understands this character, it is not merely an empirical observation that is relative to the then current state-of-the art . It is only possible to describe the behavior of things on a quantum level in a code-relative manner.[38] Attempting to understand it, to subsume it under the coherence of a subjective cogito, would be something else entirely. To have one model explaining those behaviors homogeneously would be to take and occupy the central place in the universe's transcendental domesticity. No principle can ever be fully, exhaustively, "known"—a central thought throughout Serres's books and one to which we will return. A key tenet for Serres's transcendental philosophy is that no empiricism can do without the registers of a rationalism (center, principle, foundation, conditions of possibility, etc.): these registers provide the positions that cannot be occupied immediately, but only by substitutes. They must rather be kept empty—but this *keeping empty* is an active practice. It is what the articulated zeroness of a cipher, the symbol for nought, affords. Attendant on this practice, there is a notion of criticality attached to Serres's objective transcendentalism, a criticality that depends upon the structural method that, according to Serres, affords *the importation of* contents from one domain to another (this is developed in Chapter 7, in the section "Classicism: Remembering Contemporaneity").

Zeroness. Let us recapitulate: for Serres, it is not reason as a subjective faculty that is thought to be "pure," in the sense of providing the reference ground for critique. What *is* thought to be "pure," and to provide a reference ground, is nature at large insofar as nature is universal: powerful, forceful, brilliant, violent—"*beautifully noisy*," as Serres put it.[39] It is beautifully noisy because the purity, in terms of the universe's communicational domesticity, is the "*purely multiplicitous*."[40] "La maison blanche" of the universe's domesticity takes as its ground plan an ichnography that contains the possible.[41] It is the reference for Serres's notion of criticality, and it must be thought of as an invariance, a quantitative, determinable yet indetermined and indefinitely formulatable, articulable term, like the amount of energy in the universe according to thermodynamics (as an invariant amount, energy cannot be created or destroyed, but counts as indefinite with regard to actual quantity). Pure multipliciousness can be symbolized in many ways. The architectonics of such purity and criticality is, I want to suggest, the architectonics of mathematics—the universe's transcendental domesticity is the spiritual home of all that pertains to learning (*mathemata*, Greek for "all that pertains to learning"). It is nature as indefinite but determinable, in chance-bound manner within the any-directionality of pure multiplicitousness. The elementary units in the architectonic of such a notion of nature are *pan-tomic* (*pan* from *all, all*-divisible, or divisible relative to the zeroness of the atomic, *a-tomic* literally meaning uncuttable, or indivisible). The objective representation of these units is a model of nature's *any-facetednesss*.[42] The notion of pureness thereby is the pureness of white light, the quantum matrix of color and matter. It is the pureness of a quantitative notion of mass, bare of any qualification in particular—meaning, correlatively, abundantly endowed with any possible qualification at all. Serres's domain of a transcendental objective is transcendental insofar as it issues from this inarticulate and indefinite plenty of qualifications, the nature which it grants to be addressed critically incorporates and organizes only those qualifications that can be observed *repeatedly*. Of the inarticulate and indefinite plenty of virtually possible qualifications only those that return, and to which can be attributed a certain generality, can be addressed by it. Let us remember that what the proper concern of transcendental philosophy is a *critical* manner of address. Likewise, the conception of the quantum domain in physics as the natural domain of the universe's domesticity, the home of a transcendental objective, the architecture of mathematics, can only account for things in the form of a certain generality—yet it is not the immaterial other to physical nature. There is a concreteness, an elementariness from which such generalization proceeds by abstraction: the physical *any-thingness* of white light-matter. Serres's elementariness is not the mono-valid, nor is it poly- or many-valued. It counts, in any single instant and point, as any-valued: "In short, the formal and atomic element, which is translatable everywhere within the system, turns out to be at the same time a historical condensate, which carries its roots, the law of its development and the horizon of its finality, within itself."[43]

The transcendental domain grants manners (not only ways) of addressing this indefinite and pure nature of formal and atomic elementariness in terms that are finite and determinable. This is what mathematical concepts afford. Mathematical "concepts" are symbolic, they are encrypted, and this level of algebraic (tautologous) code, spelling out zero sense or meaning, by substituting for any sense or meaning that might be made, called upon, when studying nature, is what makes the evolution of notions of sign, number, and form possible. Addressing this playground of mathematical substitutes in terms of notational partitions is indeed what makes it a *transcendental* domain (rather than a *transcendent* realm).

Equation. Let us consider now how the notion of "equation" can illuminate how to address such zeroness and the play of notational substitutes. Let us do this by reasserting the importance of finding ways of *accounting* for an impersonal cognitive agency at work within mathematics (and thereby, within computing, and within technology), one that does not separate rationality into two kinds: the subjective (human, cultural) and the objective (natural, scientific). We want to consider an impersonal subject at work amid the universality of mathematical objectivity. Our guiding question is, how to address this impersonal subject. As the cogito of an "it" (rather than an "I" or a "me"). What can we learn for this purpose from the mathematical notion of equations?

The mathematical notion of the equation was first documented in the sixteenth century, when it seems to have been introduced as what we would today call a *terminus technicus* for organizing the practice of equalizing mathematical expressions. It seems to have been introduced in European Renaissance science and philosophy together with algebra: an equation is the mathematical form for rationalizing and reasoning identity. The term "algebra" comes from the Arabic *al-dschabr*, for "the fitting together of broken parts, the forcing," and *al muqābala,* "the meeting at the opposite side." Its first appearance is usually referenced to the title of the Persian scholar al-Chwarizmi's book *Al-Kitāb al-mukhtaṣar fī ḥisāb al-ğabr wa'l-muqābala* (*The Compendious Book on Calculation by Completion and Balancing*). Two things are important to point out here: the mathematical term of an equation references a mathematical *form* for stating identity, and it does so precisely by *not* assuming identity to be given in any other way than through relations of "*appartenance*" (belonging-with) relative to an invariance (that can neither be produced nor destroyed). In this regard, it crucially differs from the identity notion in logics—it helps reason and rationalize identity, but in the original sense of Greek *mathema*, literally "that which can be learned," and *mathematics* for "all that pertains to what can be learned."

Through this emphasis on learning (rather than knowing), and hence on mathematics as an *art and techné*, the equational notion of identity is always already in a pact with the mathematical irrational (the infinitary). It remains undetermined whether identity as a postulated principle is to be regarded as a logical device, or whether there is to be assumed a substantial reality of this principle in nature that can empirically be studied in physics.[44] This indeterminateness is indeed the key aspect that Serres attributes to algebra for the advent of modern science with its paradigm of experimentation at large: "Experimentation and intervention consist in bringing it [the hidden cipher] to light,"[45] and furthermore: "At the beginning of the seventeenth century, when what came to be called the applied sciences first appeared, a theory spreads that one can find in several authors, although none of them is its sole source, which seeks to account for a harmony that is not self-evident."[46] Rather than looking to conceptual identity to furnish some self-evident harmony to be sought in nature, another notion began to spread, since Galileo Galileo and certainly René Descartes, Gottried Wilhelm Leibniz, Blaise Pascal, Bernard de Fontenelle, and also Serres; it is "the idea that nature is written in a mathematical language."[47] But Serres immediately points to the insufficiency of the term "language" here; he points out the *constitutive and active* role of algebra for the role of mathematics in experimentation. He specifies: "in fact mathematics is not a language: rather, nature is coded. The inventions of the time do not boast of having wrested nature's linguistic secret from it, but of having found the key ['chiffre'] to the cipher ['grille']. Nature is hidden behind a cipher. Mathematics is a code, and since it is not arbitrary, it is rather a cipher."[48]

Serres speaks of cipher here in a mathematical sense: cipher is a term for how, in mathematical notation, nought can be expressed. It literally meant *zero*, from the Arabic term *ṣifr*, for zero. A cipher constitutes a code that affords encryption and decryption such that once the en- and decrypting operations have been performed, the "text" or "message" ("nature," in the passage from Serres just mentioned) that those operations embrace or envelop, remains unaffected by these operations. Algebra, as the art of speculative completion and balancing, experimentally searches for the code initially without having a key. The equational notion of identity hence is capable of organizing the practice of equalizing mathematical expressions in an experimental manner:

> Now, since this idea [that the harmony to be sought is not self-evident but depends upon experiment] in fact constitutes the invention or the discovery. ... [N]ature is hidden twice. First under a cipher ["une grille"]. Then under a dexterity ["une adresse"], a modesty, a subtlety, which prevents our reading the cipher ["la grille"] even from an open book. Nature hides under a hidden cipher. Experimentation, invention, consist in making it appear.[49]

This emphasis on an equational notion of identity, whose determination correlates with its articulation in the characters of a cipher and by the rules of a code, bears two great promises: (1) It affords a thinking that is capable of leaving its object—that which it envelopes in code and makes appear—unaffected, thus giving new support to a scientific notion of objectivity; (2) This thinking proceeds algorithmically and formally, and hence can be externalized into mechanisms that perform it independently from a human cogito, but without liberating thought from the imperative to mastership and literacy. For the ability to "read" this cipher behind which "nature hides" crucially depends upon dexterity, modesty, and subtlety. In other words, a reasoning that can be externalized into a mechanism, and hence bring a subtle harmony into plain view (an interplay of parts that function well, work together fittingly, etc.), must be considered strictly decoupled from any notion of truth. In this sense, equational identity is genuinely abstract.[50]

Alfred North Whitehead's *A Treatise on Universal Algebra* from 1898[51] can be indexed as the moment in which algebraic abstractness begins to find a novel embodiment in "information." Let us take it as a guide when tracing some of the "genetic" heritage of mathematical abstraction whose lineages converge here. When Whitehead wrote his *Treatise*, algebra could only be approached by means of what he styled as "a comparative study," because it had given rise to "various Systems of Symbolic Reasoning."[52] And those systems of symbolic reasoning, as Whitehead called them, had been looked upon with some suspicion by mathematicians and logicians alike. As Whitehead puts it, "Symbolic Logic has been disowned by many logicians on the plea that its interest is mathematical, and by many mathematicians on the plea that its interest is logical."[53] This confusion—literally, materially, a confluence of "clarities"—constitutes the spectrum through which mathematics, since the early twentieth century, poses "postmodern" challenges to every philosophy concerned with separating legitimate statements from illegitimate ones.[54] Today, we encounter Whitehead's algebra (systems of symbolic reasoning) in the "artificial languages" articulated by computers—whereby calling it a "language" is to speak neither in a metaphorical nor in a clearly defined manner. The challenging question that arises from the sheer performativity of algebra's abstractness is, from a philosophical perspective, how to acknowledge that there exists something like "mathematical thinking" (against the view that mathematics "operates" "blindly," while "thinking" and "insight" pertain only to "language"). This is a challenge for the modern tradition of science, with its dualism between nature (objectivity) and culture (subjectivity); it concerns the question of how it is that mathematics "works." At stake therefore is a certain impersonal subjectivity (thought) that at once is governed by the reign of objectivity (measurement and calculation) while "instructing" this same objectivity, by developing and deliberating the terms in how it should rule.

Until the late nineteenth century, algebra was associated almost exclusively with a theory of equations, and its symbolic notion was thought to encode quantity in its classical double articulation as magnitude (metrical, answering to *how much*, presupposing a notion of unit) and multitude (countable, answering to *how many*, presupposing a notion of number). When Whitehead wrote his treatises, this had changed: with Georg Cantor's countable infinities (among many other important factors had been contributed [to name only a few] by William Rowan Hamilton, Richard Dedekind, George Boole, and Hermann Günther Grassmann), the classical distinction between multitude and magnitude gave way to a more abstract distinction between *ordinality* (answering to *the howmanieth* as in first, second, third, etc., thereby implying an ordering place-value grid) and *cardinality* (answering to *how many* as in one, two, three, etc., thereby implying the discreteness of the countable).[55] The generalized notion of quantity was now that of "sets." The status of mathematics with regard to philosophy and the natural sciences, but also with regard to linguistic form (structuralism) and "literary" form (e.g., in Ludwig Wittgenstein's notion of "natural language" as "mathematical prose") was profoundly unsettled thereby. This is what it means to say that mathematics is no longer concerned with quantity, but with symbolic systematicity.[56]

With this in mind, we can see how the genealogy of the equation has come a long way, from the practice of equalizing mathematical expressions that first gave rise to the notion, to a manner of doing arithmetic not exclusively with numbers (*arithmos*) understood in an Aristotelian sense (as ontological science), but also with what we could call "lettered/characterized numerals" (in German: *Ziffern*) that entailed an intermediary symbolic-notational formality (codes or polytomic alphabets, elements that are not nondivisible (atomic) but still partitionable in many ways). The decisive aspect is not that letters of the alphabet were newly used in mathematics;[57] rather, it is that alphabetic letters began to be used for the notation of numbers in a manner that changed the concept of number. Numbers, now, could be articulated as an interplay between variable and constant parts. This was not likewise the case in the established traditions of mathematics in antiquity. (Arguably, the meteorological as well as the divinatory traditions were concerned with something very similar, they also involved an interplay between variable parts and constant parts, some kind of proto-algebra; but it was not "numbers" that would be the subject of such articulation. Unfortunately, we cannot explore this further here.) In antiquity, numbers were determinate numbers of things, while the algebraic concept of number works upon what is a "given" only in the form of a metrical measurement point. Eventually, this novel manner of thinking about numbers gave rise to sophisticated procedures of estimation like stochastic interpolation and extrapolation. Mathematics thereby came to be seen as an activity, an intellectual and practical art, and the resulting geometry was referred to not as "elementary" (*stochastiké*, in the tradition of Euclid's *Elements*), but as "analytical," "specious" and eventually as "population based" (modern stochastics, probabilistics). The notion of a mathematical object was called by the early algebraists *la cosa*, the unknown—or not exhaustively known—"thing."[58]

This novel notion of the object triggered in philosophy (and in politics) the inception of concepts of sufficient reason, on the one hand, and of absolutism, literally meaning "unrestricted; complete, perfect"; also "not relative to something else,"[59] on the other hand. Not being relative to something else meant for mathematics that the role of proportion (A is to B as C is to D) as the classical paradigm for analysis— literally the *dissolving* (from the Greek *-lysis*, for "a loosening, setting free, releasing, dissolution"; from *lyein* "to unfasten, loose, loosen, untie") of what is *analogue* (from the Greek *analogon*; from *ana*, "up to" and *logos*, "account, ratio")—was generalized, and thereby also relativized. Proportion was now addressed rather as "proportionality," and thereby, the rationality of reason no longer only facilitates the analytic but also synthetic *manners* of reasoning. Rational reasoning was now relative to conditions of possibility and the inclinations of dispositions—and this *even within mathematics itself*. Serres's book on Leibniz, with its central notion of "Le géometral," works out precisely this important and often neglected transformation in the history of mathematics. The practice of equalizing mathematical expressions unfolds in a generalized role of proportion as proportionality, and the notion of "equation," with the symbolic forms of organizing these practices, can be understood as the technical term to express this relativization of the analogical structure of proportion. It introduced a novel art, the *ars combinatoria*, and the practice of algebraically equalizing mathematical expressions culminated with Isaac Newton's and Leibniz's infinitesimal calculus as a novel *mathesis universalis* (a universal method). Which triggered a fierce dispute in the eighteenth century between philosophical rationalism ("baroque-ish" and "orthodox" in spirit) and empiricism (reformationist and "modernist" in spirit). Kant's notion of the transcendental, together with his program of critique for philosophy, eventually calmed the disputes (temporarily).[60] Algebra, as the theory of equations, was now to provide insights not about the nature of elements immediately, but as rules that can be deduced from natural laws reigning in physics. Mechanics came to be seen as a particular case of a more general physics, including dynamics and soon thereafter also thermodynamics. It was now the formulation of these laws (no longer that of mechanical principles) that was to be stated in the form of equations, accessible critically through empirical (exact) experiments coupled with rigorous conceptual reasoning,[61] and hence decoupled from any affirmation of metaphysical (and theological) assumptions in particular.[62] But this notion of the transcendental is only capable of addressing nature "hidden under a cipher," but not its twice-hiddenness: by a cipher and "with an ingenuity, a modesty, a subtlety, that prevents our reading the cipher even from an open book."[63] If nature hides under a cipher, and, as Serres claims, "experimentation, invention, consist in making it appear,"[64] then a critical turn for philosophy depends upon being able to address this second hiding as well. It is not enough to consider mathematics as the art of learning that needs to be kept in check, critically, by empirical science. It is crucial to clarify also the knowledge that can be learned through mathematics—because it translates, and manifests, in the measurement devices applied in empirical science.

Chance-bound Objects

The main distinction to address now, with regard to this role of equations and zeroness and to the statement of identity in characteristically variable terms, concerns a danger that Serres has stated as follows: to avoid confusing invariance with identity.[65] It is Jacques Monod whom Serres had in mind with this distinction; Monod has pointed out the source of this likely confusion with regard to what he calls the "quantic revolution":[66] "The principle of identity does not belong, as a postulate, in classical physics. There it is employed only as a logical device, nothing requiring that it be taken to correspond to a substantial reality."[67] After the quantic revolution, however, the principle of identity ceases to be a merely logical device. In quantum physics,

> [one of the] root assumptions is the *absolute* identity of two atoms found in the same quantum state.[68] Whence also the absolute, non-perfectible representational value quantum theory assigns to atomic and molecular symmetries. And so today it seems that the principle of identity can no longer be confined to the status simply of a rule of logical derivation: It must be accepted as expressing, at least on the quantic scale, a substantial reality.[69]

How do we think of such "absolute identity" as constituting a substantial "reality"? For Serres, it is the identity of code and chance, an identity that is decipherable because it is chance bound, and chance bound because it is ciphered. We do have something of a vicious circle here, but this viciousness is not a negative quality. It literally means *reciprocal intensification between action and reaction,* and it is just such intensification that can be treated as formalizable and quantifiable in quantum physics. The implications of such an assumption are weighty: they demand that one come to terms with an idea of chance as an objective concept.

Serres cautions us: "It is not recommended for scientists to borrow a definition of chance from the philosophers." For philosophy, chance is either God or nothing substantial, "depending on whether the philosopher believes himself to either be God or nothing":[70]

> Before the seventeenth century, people lived fatalistically, they subjected to the stoic, Mohammedan or Christian fatum, one accepted luck and suffering as they came; one was afraid of fortune and fate, one was awed by the coincidence of events, opportunities and failings, people played dice, the goose game or solitaire, they called out to gods, themselves subject to blind powers, one died of the wind, of shipwreck; the world was a mesh of intentions, unexpected, cruel and necessary. We didn't know chance.[71]

Chance as an object was not known—"as an object, this means stripped once and for all from all traces of a subject."[72] Chance as an object was invented, he adds, by Pascal, Bernoulli, and others, for whom "tables of numbers replaced tragedy."[73] Ever since, "chance no longer has a project, but only combinations. This is, if we want, the postulate of objectivity."[74] But also since then, "this elaborate notion of chance has rather left philosophy [*déserta passablement les philosophies*], which could support it just as little as religion once did, and this is significant."[75]

Serres identifies the classical question addressed to chance as the following: Is the world saturated with bonds and connections, or are there local gaps?[76] To this question, Serres maintains, there are either "too many or too few answers"—it is "almost an antonym in the Kantian sense, for now undecidable," because "the definition is less clear than that which is to be defined: The latter affords a computation, the former hypothesis."[77] It is well possible to hold on to the former without the latter: chance can be addressed objectively through computation, but on condition that it is thought to arise from an elementary mixture rather than from an originary table of independent, individuated events.[78] In science (and this is what philosophy should learn from science),

> chance would be unthinkable—just pathetic—without an entirety [*ensemble*[79]], a basic variation [*une variété de base*]. What is important here is the plurality, the number of "factors," the pure multiplicity. Instead of the old world, where things are always already legalized, fibred into ready-made sequences, we find a table, an arbitrarily ciphered collection, a package of terms about which one makes exactly

no hypothesis at all. The thing [*la chose*] is verifiable in any place. In mathematics, such an ensemble is called a collective.[80]

But how could philosophy make sense of such a notion of the "collective," a "pure multiplicity" that counts a magnitude that is, so to speak, "intra-mathematical," referring to a "population" of instances that are not, somehow, rooted in a hypothetical nature like that of numbers, or physical forces, or metaphysical ones like a life force or, indeed, fate and predicament. As some form of original plan or prime judgement (which would imply that the world is no longer always already *légalisé*). It is here that we encounter what is at stake with a notion of nature that hides twice, behind a cipher and behind dexterity. Pure chance, from such a perspective, is not chance exposed *from* all cover-ups, not chance *laid bare*, naked and without disguise. The mathematical theory of information is capable of addressing chance *in* its covering cloaks—we are dealing here, according to Serres and the role that he attributes to chance, with a notion of chance that is substantial not despite but *due to the fact* that it only appears masked, objectively so, *covered* by code. Pure chance is chance stripped bare of all hypotheses; it is chance in purely quantitative terms—but "purely" in relation to "quantitative" means (counterintuitively so) "ciphered," *rendered countable by code*. The objective appearance of chance which Serres foregrounds is its "signaletic" appearance, because signals (in electro-engineering) can only be measured relative to a signal horizon, a background against which they appear as measurable. It is this that is constitutive for their signification as signals. And this background is not that of a decorative order (the literal meaning of a cosmos), but one against which chance appears like clouds in the sky: "This background ... may be immense, but it is no longer dominated"[81]—it is in this sense that Serres calls this background "chaos." As such it is no longer dominated; it is chaos rather than a god or nothing.[82]

How can chaos, as "the background" before which objective chance "appears like clouds in the sky," be referred to with a definite article (*the* background)— without saying thereby that this referential "chaos" is, somehow, dominated? This is the core of Serres's conception of "objective chance" as chance that remains mysterious, yet is computable. The main inversion here is to consider code not as rules that are derivative from some law. Or, rather, code is to be considered as quasi-rules, rules that are derived from the only quasi-principle: the principle of chance, the only principle that is not bound to make any exclusions at all. No hypotheses need be ruled out when adopting chance as a principle. But how then can chance be "definite," "objectively determinable"? How can there be an objectivity to it at all? "Chance would be unthinkable—just pathetic—without an ensemble, a basic variation [*une variété de base*]."[83] Chance can be considered as computable and as objective because it presents itself masked by *informational* code. Informational code complements it with a "basic variation," rather than with a "foundation" in the analytical sense of the word "foundation." Code as signaletic appearance that raises chance up against the background of chaos. Code allows for *indexical* reference, determination that is purely quantitative. Code, as that which makes chance appear objectively, affords neither signification nor the expression of something meaningful—it *indicates* chaos as the negative to its own, any-valued, vicarious *orderality*: code is "positive" in the same sense as substitute operators *as* substitute operators are positive, that is, with regard to the means of transcription only, not any content of a "script." It is the *quasi*-positivity of a spelled-out zeroness, the totality of operations afforded by an alphabet of code that spells out a particular cipher, and hence indexes "nothing" without substantializing this "nothing."

Taking Ignorance into Account: Quantifying Strangeness

The world lost its mute status of being the objective set of passive objects of appropriation so as to again become what it had never stopped being, the universal hotel or host for inert things, for living species and ours in particular, thinking, active, suddenly knowledgeable and strong enough to speak equal to equal with it. And it responded to us like a quasi-subject.

Michel Serres, 'Vie, information, deuxieme principe'[84]

Entropy and Negentropy

"Thought interferes with the probability of events, and, in the long run, therefore, with entropy":[85] the term "negentropy," or negative entropy, is born from this very situation. By addressing the term like this, in relation to thought rather than to the predominant association it is usually referred to, I present an unusual angle for thinking about these two concepts, entropy and negentropy. The usual association of the term is the greater capacity of living organisms to counter a certain susceptibility of physical systems to degrade in a long-term tendency to states of equilibrium. Entropy is a key term in thermodynamics, and it's negative, negentropy, was introduced by Erwin Schrödinger to distinguish biological systems from physical systems.[86] It was intended to mark the greater capacity of metabolizing, living organisms to maintain, consolidate, and intergenerationally to increase the stability, complexity and adaptability of their species' orders. The reason for choosing, as I do here, the unusual angle which relates the concept of "negentropy" to "thought" rather than preserving it as a mark of distinction between living and nonliving systems is perhaps not entirely obvious. The main reason for doing so is that Serres's interest in negentropy is clearly referenced to the molecular biologist Jacques Monod, and his insistence that from a quantum physics point of view, identity can no longer count as a logical device only, rather, as elaborated earlier, it needs to be attributed a substantial reality. Departing from this, Monod's concern had been how to avoid confusing such identity with a notion of invariance; invariance was well known from thermodynamics well before the concept of "negentropy" was coined; thus I want to reformulate the concern here as: How can we avoid confusing this substantial notion of identity after the quantic revolution with invariance in the thermodynamic terms of entropy? Serres adapts this concern regarding the distinction between invariance and identity entirely from Monod; for his unique approach in thinking of communication philosophically, but in a manner that maintains compatibility with quantum physics, it is key to understand that an ontological distinction between living and nonliving, between organisms capable of biological metabolism and physical systems that are not capable of it in the same sense, can no longer be a relevant focus point. Rather, for Serres there is a universality to nature that applies to all existing things, living or not; it consists in sending, receiving, storing, and processing information. This is how communication (through information theory) must count, for Serres, as *physics*. As did Monod (I will come back to this at greater length shortly), Serres affirms a stance according to which "life" figures among the phenomena of "physics" in the sense of a natural philosophy, not physics as an academic discipline that has to keep clear distinctions to chemistry, for example, or biology. It is against this background that the current chapter attends to the concepts of entropy and negentropy within an angle that relates them both to "thought" and "communication," rather than to "metabolism" and "ontology."

We saw how negentropy was introduced by Schrödinger to distinguish biological systems from physical systems. Let us see now how it can be related to thought and communication. Shortly after its introduction, the concept of negative entropy was picked up by Léon Brillouin, another quantum physicist. He generalized it in order to apply it critically within the domain of Claude Shannon and Warren Weaver's mathematical communication.[87] With this, negentropy became a key concept in information theory, and in how information theory began to play a bridging role between physics, chemistry, and biology within what came to be called "molecular biology."[88] This was arguably the key development in twentieth-century science that was to trigger an academic battle that is still ongoing today, over which of the two disciplines, chemistry or physics, was to be "in charge" over the question of life; or, in short, over the domain proper to biology. The perspective from which negentropy will be discussed here situates the term amid a particular problematic, which is the following: What role does probabilisitics play for the central paradigm of empirical experiments for science? If our interest in negentropy was one couched in terms of the natural system's capacities to metabolize (rather than one in terms of natural communication, or even "thought" as a natural capacity, as it is for Serres, and as I attend to it here), then the proper question to ask regarding negentropy would be something like: "How to think about the proper ontological status of the objects of empirical science?," or even "How to think about the proper ontological status of probabilities and chance in the experimental paradigm of empirical science?." But by regarding communication as a universal given that applies to all things natural as well as so-called cultural, to all things living and nonliving, as it is crucial for Serres's approach, this emphasis on ontology, on how to classify things, cannot be the adequate angle to pursue. Rather, if communication is universal, then the angle to pursue ought to be a categorial one (not classes, as criteria for distinctions); because categories are related to principles (and hence either to law or metaphysics), they organize the "attributes/properties" that belong to every existing thing, while classes are related to an empirical science and organize order in terms of kind, genus, and species. This is why the preceding chapter has introduced Serres's notion of how we can adopt chance in science as the sole principle that needs not rule out any particular hypothesis at all, and hence does not compete with or even contradict empirical approaches. Objective chance is the sole principle that can always be applied objectively and rigorously, if profiled against the background of nature as chaotic—a background that with the rigorous notion of "objective chance" is no longer dominated; a background that no longer figures as a *substantial* nothing or as god, if only we respect that chance can be treated quantitatively (not qualitatively). We saw how chance can be considered as computable and as objective because it presents itself *masked* by *informational* code. Code allows for an *indexical* reference, determination that is not concerned with the contents of signification. What we need to better understand now is how such *quantization* facilitates the *quantification* of objective chance. This is what I propose to call in this chapter "the quantification of strangeness," and the crucial component for it is provided by means of information theory.

So let us begin again from where we started. We set out from the idea: "Thought interferes with the probability of events, and, in the long run, therefore, with entropy,"[89] and began to ask about how such natural communication (information as a physical quantity) impacts our current paradigm of empirical experiments for science at large. What this impact consists of in short: the centrality of the empirical paradigm needs to come to terms with how the algebraic encryption of quanta plays a role in measurement, and interferes there with the non-probabilistic practices of measuring and counting that were constitutive for classical physics. Probabilistic procedures are applicable only by means of computations performed upon ciphers, the mathematical way of articulating *nought*, *nothing*, and the equational manners of balancing and completing by involving an encrypted negativity to all countable and measurable positivity: for probabilistic reasoning, the total number of possible cases, the total range of possible outcomes, each with its determinable likeliness of occurrence, must be finite and countable. This is no less true for negentropic than it is for entropic processes, though negentropy accordingly quantifies and makes countable the symmetrical negative of what the term "entropy" quantifies and makes countable.

Let us remember how the term "entropy" was introduced by Robert Clausius. It was in the want of "a word for measuring a quantity that is related to energy, but that is not energy."[90] The idea that there is any direct interplay between thought and the particular quantity, entropy, was a radical one. Its radical nature must be understood in the context of the dominant paradigm at the time when it was introduced. The modern scientific assumption at this time was that thought cannot affect the nature of its object, only the subjective understanding thereof. While subjectivity depends upon will or intent, natural forces work determinately and gratuitously. We can formulate the analogue of this situation in terms couched in thermodynamics, insofar as the conversion of heat into energy, or energy into work, leaves the amount of heat unaffected (in this analogy, heat represents the nature of thought's object). The modern assumption translates here as: the total amount of heat in a system remains constant; it merely passes from a hotter body to a colder one. At the same time, the second law of thermodynamics adds a contrast to this principle. It states that we cannot maintain an identity between heat and energy: "No matter how much energy a closed system contains, when everything is the same temperature, no work can be done."[91] It is the *unavailability* of this energy that Clausius wanted to measure.[92] If heat is regarded as the manifestation of energy in a system, what is needed is a distinction between energy that is available for work, and energy that is not available for work: the total amount of heat in a system may well be constant, but it cannot be transmitted (from a warmer to a colder body) *without some work taking place.* The work done by natural forces in the thermodynamic setup can therefore not be regarded, after all, as "gratuitous" in the same manner as it is in classical physics. The notion of entropy hence introduces a distinction between "free" and "bound" energy through which thermodynamic experiments can be constructed as so-called thought experiments, whereby the formulation of the experiment in question—a mere exercise in thought—appears *directly* to affect the laws of nature. Before attending to how the introduction of an analog distinction with regard to negentropy, as "free" and "bound" information, in negative symmetry to "free" and "bound" energy, will lend itself as helpful in *mediating* this apparently "immediate affection" of natural laws by thought exercises, let us look more closely at what is at stake with the entropy-related distinction of "free" and "bound" energy.

Thermodynamic processes introduce a certain irreversibility into how we think about the conversion of energy from one form to another—"Time flows on and never comes back," said Brillouin in 1948.[93] This was consciously to contradict the classical formulation of natural laws, which are understood to apply with indifference to the arrow of time. The angle of incidence of two colliding bodies will calculably dictate the angle at which they depart each other, for instance. This was held to be as true under a reverse arrow of time as for a forward one. This indifference is classically assumed true of all natural phenomena. When the modern paradigm for experimental science builds on the assumption that thought leaves untouched the natural object it tries to conceive of, we can see now that it is exactly this assumption which appears to break down with the new kind of physics (quantum, information).

But a word of caution first. The irreversibility of thermodynamic processes cannot straightforwardly be squared with the indifference of natural laws to the arrow of time; the energetic state of a system evolving into the future should be expected not to differ from that same system treated as "evolving" into the past. This very likely not only demarcates the limits of chemistry in accounting for phenomena of life, it is also what probabilistic reasoning (as against stochastic and statistic methods) challenges. Probabilistics is algebraic; its code-boundness, and its irreducibility to either stochastics or statistics are key factors that characterize the "quantic turn" in physics. As Brillouin explains: In the measurement of any physical system, there are macroscopic and microscopic variables to be taken into account. The former refers to those quantities that can be measured in the laboratory, but they do not suffice to define completely the state of a system under consideration. Once a system is also considered in quantum terms of its radiation and absorption, there are an enormously large number of microscopic variables to be taken into account as well—and these, one is unable to measure with accuracy with regard to the positions and velocities of all the individual atoms, quantum states of these atoms or of the molecular structures, and so on. "Radiation is emitted when a physical system loses energy," Brillouin explains, "and absorbed when the system gains energy."[94]

But before attending to the new kind of physics (quantum, information), let us stay a bit longer with thermodynamics. In thermodynamic processes, energy is not lost; rather, it dissolves. It becomes "useless." This is the so-called expense problem related to the irreversibility that applies to thermodynamics. The total amount of entropy (unavailability of energy for work) in all physical systems that can be studied empirically, experimentally, necessarily seems to increase. There is hence a source of disorder that applies to systems which "seemed strangely unphysical," that even "implied that a part of the equation must be something like knowledge, or intelligence, or judgement," as James Gleick puts it in his recent study *Information: A Theory, A History, A Flood* (2011). He continues, "Dissipated energy is energy we cannot lay hold of and direct at pleasure, such as the confused agitation of molecules which we call heat."[95] At the time when Schrödinger and Brillouin were developing their notions of negative entropy, it became clear that heat cannot be regarded as either a force or as a substance. Heat was not to be regarded as an equivalent to energy. In the course of these developments, order—as the epitome of objectivity—acquired a certain amount of subjectivity. In a certain sense, the eye of an objective observer appears inextricable from the thermodynamic notion of order: "It seemed impossible to talk about order or disorder without involving an agent or an observer – without talking about the mind."[96] For these reasons, the formulation of the second law met with controversy: it seemed at once to capture a plain, demonstrable fact of nature which nevertheless seemed to originate only with human design, through human thought. It is based entirely on observation. Its philosophical or even cosmological implications, if it indeed is a law, are immense: it renders inevitable (however distant) the fatality of all life on earth. Lord Kelvin, a mathematical physicist and engineer who did foundational work in the mathematical analysis of electricity and formulation of the first and second laws of thermodynamics, was not the only well-established scientist who began to consider the consequences for science of this "heat death" (as it became known) of the universe. Resolutions to this problem began to be discussed in terms of the possibility of a perpetual kind of motion that began to be linked up with the conjecture of "a perfect experiment": one that would be liberated from and independent of the biased reasoning of imperfect human faculties and their limitations. This ideal arguably still haunts today's discourse on artificial intelligence.[97]

Let us jump now to the introduction of the term "negative entropy" more directly. Schrödinger introduced it in *What Is Life? Mind and Matter* (1944) as a term that expands the thermodynamic view from physics to biology. Such a generalization promised to relativize the implications of the physical view on the entropic universe (heat death). Schrödinger's point of departure is the observed capacity of animate systems to metabolize—by incorporating and binding energy into various scales or "cycles," through processing it into "nourishment" and "waste" (the terms are relative to a particular species, they are not absolute; what is waste for one species may well be nourishment for another). This energy that feeds metabolism, he called "free" in the sense of "available," or "unbound." Negentropy came to be, for Schrödinger, a term that can quantify life (while not being life) in the same way that for Clausius, entropy could quantify energy (without being energy). What had been the energy-expense problem of work for James Clerk Maxwell (whose thought experiment on a thermodynamic "demon" we will attend to shortly) turned henceforth into a veritable economy in terms of import and export at work in the biosphere world of thermodynamics. Organisms import negentropy (quanta of life), as Schrödinger put it, and the more they do so, the more they rid themselves of entropy (quanta of physical entropy now conceived as disorder, relative to an organism's [a species'] metabolic and temporary order/organization). Accordingly, the opposition between thermodynamic entropy and biological negentropy assumes a much stronger correlation with the opposition between "order" and "disorder"; but it lacked a means to keep apart the different scales that would facilitate addressing the biosphere world in its full complexity that comprises living organisms as well as nonliving systems like minerals, or atmospheres. In this contrast between classical physics and thermodynamics-based biology, the biological paradigm now seems to contradict the second law of thermodynamics, and instead suggests that the metabolisms making up the biosphere are in fact capable of also decreasing, rather than only increasing the universe's entropy (the amount of work unavailable in the thermodynamic universe). The competing paradigms contrast like this: while thermodynamic physics is the entirety of physical phenomena in the universe (as, ultimately, one generic nature), biology treats only the specific natures of life forms. The physicalist notion of entropy, which in physics started out as denoting not *the absence of order* but *the virtual presence of order in any of its possible variations*, comes to represent, in light of how biology's operational term of negative entropy can quantify life according to specific forms, a critique of universalist thought in science. It demarcates the *relative* absence of possible variations of order on the scales of particular ontologies; it is through such thinking, across heterogeneous scales, that entropy came to mean the *relative* absence of order, or, in short, a peculiar measure of "disorder." But "order" and "disorder" figure now, according to such measurement, on heterogenous scales that are inverse to each other. Hence, this important critique against a certain hegemonic universalism remains without proper categories that could provide a "transversal con-versation" ranging across the different ontologies. This lack still informs the current ecological paradigms and their central notion of how to think about *the climate*—in effect, it threatens to entirely abolish the scope of universality for science. This would abolish the distinction between scientific knowledge and the knowledge provided by traditions and cults, it would subject science to a condition of "cultural wars" not unlike those of "religious wars" that have always characterized history. The guiding question for our pansemiotics' view, provided by our postulated "quantum literacy," is how the scope of universality could be saved for scientific practices, to continue setting them apart from particular cults and cultures; distinguishing information and code from both cultural techniques, form and number (mathematics) as well as signs and linguistics (languages), offers an outlook of how this could perhaps be achieved.

The introduction of "information" into the thinking about thermodynamic processes managed to abstract from, and to open up, this troubling impasse between a certain monism (the entropy of a sole universe) and its pluralist counterpoint (the entropies of different species). I can only point briefly here to how this "conversion" between information and energy works. A key reference for it is quantum physicist Léon Brillouin's adoption of Schrödinger's term "negative entropy" in a manner that introduces an algebraically quantized notion of information as a contracting bridge between the different notions of entropy in physics and biology. Brillouin conceives of information as a kind of currency that circulates in energetic expenditure (the import and export between systems) on quantum scales that apply also, immanently, to macroscopic scales. Such that "all these macroscopic and microscopic variables of unknown quantities [of which he speaks in the citation above] make it possible for the system to take a large variety of quantized structures, the so-called Planck's complexions."[98] With this, Brillouin postulates information science as the proper domain for quantizing the relations between physical entropy (the virtual presence of any order) and biological entropy (the absence of order, disorder). It can provide for such relation without subjecting one to the other. Familiar with the work of Alan Turing,[99] Claude Shannon[100] and Norbert Wiener[101] on a mathematical notion of information, and their dispute on whether information could be measured in terms of the experimental notion of entropy applied to physical systems (Shannon), or whether the better model is Schrödinger's terms of negentropy—as "relatively entropic domains" for import in biological systems,[102] Brillouin foregrounded the role of "code" in such "intelligent" computation: he thereby applied a *double* notion of negentropy and entropy—one to energy, one to information. Entropy applied to *energy* follows a physicalist view (universal nature), negentropy applied to *energy* follows a biologist view (pluralist natures, fragile balances of import/export relations in metabolisms); entropy applied to *information* follows a biologist view (attends now to specific balances of pluralist natures, whereby local balances are being "generalized" into global "ontologies"), negentropy applied to *information* follows a physicalist view (universal nature, can now attend to that nature's "givenness" in greatest local diversity ("givenness" as "datedness"; it will be elaborated in Chapter 3)). The fourfold of energy and information orders assumes that both are linked, and are convertible to each other, by code. Such convertibility made it possible to coin the symmetrical couple of positive and negative entropy in terms of "free" and "bound" relative to code (as the currency of such convertibility); like this, "free" and "bound" no longer characterize particular world views (physicalist universalism vs. ecological pluralism), but rather how the two perspectives on the world inevitably need to "contract" one another: pluralism needs a monism to distinguish itself from, and monism needs pluralisms to integrate and accommodate. The qualifiers of "free" and "bound," with regard information and energy according to Brillouin's quantum physics view, apply to nature via expenditure as an economy, but as an inextricably *double* economy—not unlike the practice of double-entry bookkeeping, which emerged with Renaissance algebra. The nature of such a communicational physics is economical, but doubly so—once in thought (when applied to information) and once in manifestation (when applied to energy). Brillouin called entropic information "free" information, and referred it to the maximum number of permutations available in any code, through exhaustive recombination of all its elements (code thereby means any finite system of ordered elements, like the Morse code, or the Roman alphabet, the periodic table in chemistry or the DNA in molecular biology). These a priori cases can be computed by combinatorics. Such entropic information is called "free" because each of the possible combinations must be regarded as equally likely to be actualized. In other words: free (entropic) information coextends with the total range of articulation for the zeros of a cipher. It is free, but still it is relative to the code in which it is communicated. Bound information, in contrast, is referred by Brillouin to negentropic information. It is empirically measured information (in experiments, where the particular measurement afforded by a particular code manifests itself). This inclination in the measurement of information allows for thinking of information via a kind of currency (code)—an operator capable of establishing general equivalence, equivalence between observation (bound, negentropic information) and object (free, entropic information). Information conceived thus is thought to circulate in the physical expenditure of energy in the work executed, as well as in the economy of import and export in a biological system's metabolism. "We cannot get anything for nothing, not even an observation," states Brillouin, citing Dennis Gabor.[103] This very important law is a direct result of our general principle of negentropy of information, Brillouin elaborates. When Brillouin speaks of his "general principle of negentropy," he speaks of a manner of generalization achieved by regarding code as the convertible operator, the currency at work in expenditure. It is an *abstract* generalization. He goes on to underline that, "it is very surprising that such a general law escaped attention until very recently."[104] The acquisition of information in measurement not only has a price, but also yields something: an increase in operational power. This is an idea that would contribute to the development of a theory of how to quantize (mathematical physics), and hence also quantify (logics, epistemology) both energy (Clausius) and life (Schrödinger).

This law, that of the Price of Information, applies to the development of artificial intelligence too (like Maxwell's ideal thought experiment)—but the artificiality in question here does not exclude the naturalness (physicality) of this intelligence, because it is a theory on the "natural powers of abstraction."[105] This very idea, that information and energy articulate each other in a natural and evolutionary, "communicative" dynamics, can account for such natural communication to be engendering, in mutually reciprocating ways, contractual (code-relative) powers of abstraction that can be specified, and to which both submit. This renders the implication of a perpetual motion unnecessary. The threat to the experimental paradigm in science is thereby averted. We saw that the second law of thermodynamics suggested that at one point the universe would no longer be able to perform work because all energy will ultimately be rendered useless; we have also seen how Schrödinger's generalization of thermodynamics to biology was capable of relaxing this outlook, by demonstrating how, through the metabolism of life forms, entropic dissolution is not the only trend of irreversible time. All life forms organize negative entropy in their organicity and the domain of life, in its richness, withstands death in its universality. With Brillouin's further generalization of negative entropy in biology to quantum physical information theory, life and death can be quantized, and quantified, in ways that affords *dating* the objects of science in terms of a Chronopedia. In Chapters 3 and 4, "Chronopedia I: Counting Time" and "Chronopedia II: Treasuring Time," I will describe this form of organizing knowledge as an index of the universe in terms of its overall contemporaneity, and speak of the *agedness* of the universe. Such *dating* is something Serres develops further, especially in *The Incandescent*, where he couples Darwinian evolution with an inverse which he calls Exo-Darwinian evolution. What I call here "powers of abstraction" is described by Serres, with reference to Monod, as different scalar *niveaux* (levels) of entropy and negentropy. We will come back to this in more detail in the following chapters.[106] What Monod realized, according to Serres, is that with Brillouin's theory of information, information can be transformed into energy (e.g., as electric current) and vice versa; energy can be transformed into information (through studying distributions of heat in terms of metabolisms) without reverting to any form of animism or spiritualism.[107] It is important to note that Monod's book has received a great deal of attention since its first appearance—though not always sensitively. Serres's reading of Monod demonstrates extensively that it is precisely an affirmation of the mediating and non-neutral role of code that makes Monod's stereometry so valuable and prescient, while this very aspect, this non-neutrality of code, is what has been largely ignored by the scientific community. So much so that the name Monod stands today—quite inversely to his stated position—for a style of mechanist views on organisms that seems to be at odds with evolutionary approaches in biology via morphogenesis. I cannot discuss this adequately here, but it needs to be stated clearly to prevent confusion. It would be helpful to distinguish this abstract and operational aspect of code with regard to DNA as a naturalized reference for genetic code; we should speak of the theory of information in biology as a *quantum-crypto-graphic* theory. It is a theory that lends itself to a geometry of spectra in order to distinguish the different *niveaux* of entropy and negentropy.[108]

But how can we imagine information and energy achieving, in their interplay, powers of abstraction as different scalar *niveaux* of entropy and negentropy that can be distinguished objectively and specifically? What Brillouin calls the Price of Information is not merely a metaphor for the idea that thought affects its object.[109] In mathematical physics, the Price of Information can be quantified precisely; it can be indexed even with a number (10^{-16} in Brillouin's 1956 book, a number which by the state-of-the-art particle physics of today has reached 10^{-32}).[110] The rising value of this coefficient refers to the increasingly small scales at which nuclear science is able empirically to observe the behavior of particles. In the case of 10^{-32} it is the coefficient allowing the famous Higgs Boson to be traced in the CERN Accelerator. What is at stake with the Price of Information from an economic point of view (there is a properly philosophical point of view, too, which to Serres is the decisive one, and to which I will turn at some length in a moment) is the large amount of energy needed to facilitate experiments on such tiny scales. The enterprise of physics is subject to the what is called the law of diminishing returns. While the Price of Information can be generously ignored for all experiments other than those on a quantum scale, the energy needed to extract information at ever smaller quantum scales escalates disproportionately for obtaining speculative and uncertain results: so too do the costs. By the law of diminishing returns, the acquisition of the tiniest bits of information comes at the greatest costs. The disproportion at stake is not primarily one of economic decisions (although it is this too). But we can easily see in those accelerators a kind of contemporary cathedral—cathedrals of science— and who would deny that such monuments are crucial for the building of cultural identity, particularly in the current period of a supposedly secular political praxis? It is part of the power of monuments to be incredibly expensive in every possible aspect.[111] But financial prudence is not Serres's main concern here, nor is it ours.

What, then, is at stake?

What is indeed disconcerting about this theorem of the Price of Information, is its unproblematized coexistence with the law of diminishing returns. This coexistence remains unproblematized as long as the central assumption is that of an ideal equilibrium, that usually serves to ground non-information theoretic epistemologies. This is disconcerting because doing so continues to exclude contingency, as objective chance, from science. Such epistemologies thereby inevitably purify and idealize the object of science. Chance and necessity cannot easily be reconciled without exposing the object of science (or philosophy or politics or economics) to a certain openness for the irrational. Serres elaborates in *Statues* (1987):

> Enlightenment philosophy teaches that the irrational must be driven out: What do the hideous statue and its inhuman form of worship have to do with us? But we have since learned to call anthropology what the Enlightenment cast out as madness or darkness, and we have also learned that exclusion brings us back to the sacred because the gesture of expulsion precisely characterizes sacrifice. By rejecting this form of worship and scene as barbarous, we risk behaving the way the Ancients did. Therefore let's accept our anthropological past as such; ignoring it would make it return without our suspecting it.[112]

The Price of Information as a Measure for an Object's Strangeness

We have already briefly touched upon the idea of addressing objects in their strangeness early on in the chapter, with the treatment of equational thinking, ciphers and what we called their articulation of zeroness. Only if we avoid confusing this novel notion of identity after the quantic turn (identity as substantial, and not merely as a logical device) with invariance, are we able to appreciate the strangeness of objects in all mathematical formality.[113] Let us look closer now at this distinction, for which Serres's key source is Monod's early and groundbreaking book entitled *Chance and Necessity: On the Natural Philosophy of Modern Biology* (1970), where he introduced Brillouin's generalized information theory (which is, as discussed earlier, crucially different from Shannon's in that it gives a central role to code) into molecular biology.

Monod approached Serres after one of his lectures and timidly asked for his opinion on a book he had been writing, which seemed to him to have great impact for philosophy, but which he had written without being a philosopher himself. It turned out to be the manuscript of *Chance and Necessity*, and it was to contribute significantly to the growing discord between the young Serres and his teachers, Georges Canguilhelm and Gaston Bachelard.[114] The point of discord revolved around their respective notions of complexity, with their respective ethical, ideological, and certainly political, implications. The interest they all shared was how to accommodate contingency, and hence the role of chance, within the emerging techno-scientific paradigm. What Monod proposed in his book is that chance, as the scattering of direction, must be attributed a substantial role from a scientific point of view, and this without it eliding the substantial role necessity (and the directionality it entails), must/should play for method. The tools that allowed him to produce and develop these propositions rigorously are those of information science (crucially so the work of the above-mentioned Brillouin), as well as those concepts linking thermodynamics to life as advanced by Erwin Schrödinger among others.

Monod proposed addressing necessity as "teleonomy," and chance as "invariance." Invariance, in this sense, is the code pool of all the genomes that exist, across all beings, and hence pertains to the level of the genotype. Teleonomy refers to the level of the phenotype, and applies to structures of speciation. The elements that scandalized many were first, the claim that invariance preceded teleonomy in biochemistry, and second, that this is generalizable to "code-pool" and "structure" such that it is applicable to *any* object that science is concerned with. This generalization was indeed the reason why Monod considered it a philosophical book—the prolegomenon for a new natural philosophy, as Serres calls it—rather than a disciplinary book for biochemists.[115]

There is a universal nature (which Monod calls "substantial identity") that pertains to all things, and this universal nature is what Monod suggests addressing in terms of an object's "strangeness."[116] One confuses invariance with identity whenever one reads Monod's use of the words "strange" in his discussion of "strange objects" as an adjective, Serres elaborates: "strange does not qualify a substantive"; rather, "strange" operates as a quantifier, not as a qualifier. The molecular theory of the genetic code is a physical theory of heritage that complements the Darwinian evolutionary view. A physical theory of inheritance is the basis according to which substantives are quantified in terms of different *niveaux* of negentropy (as amounts of chance that can be deciphered from the mutually implicative relation between invariant and teleonomic information). With his notion of the *strange object*, Monod offers a notion of the object that neither contradicts the principles of physics (second law of thermodynamics, one universality) nor those of Darwinian biology (natural selection in evolution, pluralist universality of natural kinds). Hence ontogenetic life and phylogenetic life are both compatible with thermodynamics. But the central assumption thereby is that invariance genetically, chemically and physically needs to be considered as preceding teleonomy: "the *genomenon* is the coded secret [*secret codé*] of the *phenomenon*," as Serres puts it.[117] The "strange object" is one which neither presupposes a distinction between natural and artificial—things endowed with a purpose (project) and things natural (without purpose)—nor one between animate and inert. One concept in particular is crucial for understanding the primacy Monod accords to code; a concept that has often been misconstrued under the gloss "the code of life." This is, for Monod, not a genetic code as such, but a relation between code and its secret (life) that is one of parallelism in structure, one regarding the phenomenon of life (not life itself) and one entirely chance-bound in the substantial definition of the term:

> Between the occurrences that can provoke or permit an error in the *replication* of the genetic message and its functional consequences there is also a complete independence. The functional effect depends upon the structure, upon the actual role of the modified protein, upon the interactions it ensures, upon the reactions it catalyses—all things which have nothing to do with the mutational event itself nor with its immediate or remote causes, regardless of the nature, whether deterministic or not, of those causes.[118]

In order to illustrate this postulated independence between the code itself and its articulated manifestations, Monod discusses the coincidence of genuinely heterogeneous sequences using the example of a worker fixing a roof, letting go of his instrument accidentally, and a passerby, who is in no way related to the worker on the roof but, who is hit by the falling instrument and killed. If one were to assume a larger logic that homogenizes these two heterogeneous series, one would indeed have to assume that the passerby was fatefully predicated to die like this from the very beginning of his existence. Against the assumption of such fatalism, Monod stresses his twofold definition of chance, as: (1) operable; and (2) substantial. While the genetic text itself is a closed and finite system (with its residual alphabet of amino acids and nucleotides), the source of the biosphere's incredible variety results from errors in the transcription of the code's sequences. For this transcription process, *pairs* of nucleotides (literally *it is a pair which "contains" a nucleus*) are indispensable—single nucleotides will not do. These pairs engage in a transcription process and here lies the essential role our introduced distinction between entropic and negentropic information (as a double pole vis- à-vis that of entropic and negentropic energy), plays: invariant amounts of information (two versions of free, entropic information, twice the alphabet (as cipher) of the "genetic text") counterbalance the structural teleonomy of the process of binding negentropic information (manifesting the biospheres' variety). This process of transcription must be deciphered in a double sense, like a Rosetta Stone. It is what Serres means when he says: "nature is hidden twice. First, by the cipher. Then with an ingenuity, a modesty, a subtlety, that prevents our reading the cipher even from an open book. Nature hides beneath a hidden cipher. Experimentation and intervention consist in bringing it to light. They are, quite literally, simulations of dissimulation."[119]

Let us articulate in a crystalline and formulaic manner some helpful symmetries that connect what we have discussed thus far:

1. Invariance is the term reserved by Monod for a quantity that provides the source for *niveaux* of integration[120] (through catalysis of free and bound information).
2. Invariant content establishes what Monod calls "teleonomic information." Teleonomy refers to probabilistic calculations on the combinatorial total of transfers that can apply, in conformity with the laws of conservation, to an invariant amount of information.[121]
3. It is Monod's great achievement to have distinguished a twofold notion of chance: (a) an *operable* definition of chance as the *unknown* (due to the imperfect experiment, not knowing all the initial conditions, the cost of observations), that is, chance at work in logical/formal terms in stochastic statistics; and (b) an *essential* definition of chance, which attributes to chance a substantial and absolute identity.[122]
4. We can readily associate *operable chance* with *entropy as an operational measure in thermodynamics*, and *essential chance* with the *principal assumption that the amount total of energy in the universe be finite and invariant.*
5. The latter (total amount of energy in the universe) cannot be counted. It can only be coded. The symmetry here is to think of universal identity as a code pool that indexes entropically all life forms in terms of DNA. Like the total amount of energy, it cannot be counted, but it can be encrypted and deciphered by translating ("transcribing") between manners of coding in the terms of its residual alphabet of amino acids and nucleotides.

What this precedent of invariance (a code pool) over teleonomy (a structure) means, in other words, is that a scientific method depends upon affirming—without having a particular framework to support such affirmation—*that* code constitutes the methods for studying phenomena, just as physics accepts *that* mathematics constitutes the methods for working empirically. Readers familiar with Serres, especially his books on Leibniz (the relation between the system and its models, between the *ichnography* that contains the possible and the *scenographies*)[123] and Lucretius (the *clinamen*, as a ciphered atom letter and angle of rotation that is minimal in a structural sense, that is, minimal in myriads of ways, every time relative to each local situation),[124] will easily recognize this idea. In other words, what Monod set out to establish, explicitly for biochemistry but implicitly also for physics and philosophy, is that, from a scientific point of view, necessity *originates* in chance. Natural laws are derivable from the aleatory play of pure multiplicities, the only law that is not bound to exclude any hypothesis, and hence the only law that is not a positive metaphysical principle or a divine law. With regard to this aleatory play of *pure multiplicité* he elaborates:

> Is it so surprising that it stems from pure multiplicity? Chance is not something and it has zero dimension, just like heat and information. Think of it as a thing and give it dimensions, and you will find a bound world [*un monde lié*] or a God above the Gods, that is to say, necessity, its opposite: an anti-physics, a metaphysics. No, chance is number, a play of numbers. It is even written in numbers.[125]

But what does this mean? Nothing more and nothing less than that the most improbable phenomena, which we customarily feel justified in neglecting by the paradigm of the old bound world, always already ordered into and fibered with ready-made sequences, are in fact rather strong and forceful indicators: "The most improbable phenomena, when they exist, reproduce. They appear [*se produise*] because they reproduce [*se reproduisent*]."[126] Order can no longer be considered the norm—it is the exception to the rule, as Serres is not afraid to say plainly:

> The physicist has become anarchist, his domain is the background noise [*le bruit de fond*], and music is rare. Every object must be called a miracle. The astonishment of old has overturned the roles. We must accept this overturning, for which the Copernican revolution is child's play, in order to enter science today. That it be unacceptable to our cultural stereotypes and the taste we've acquired in how we live in our own homes, is obvious. And yet, it turns [reference to Galileo]. This translates today to: And yet, it is extraordinary that it turns. Law no longer resides in law, and mankind is no longer in the world. It is incomprehensible that there are things which are comprehensible. And yet, there is order. Conservatories of chance and auto-regulated systems, packed to the gills with negentropy.[127]

It is clear that there are entailments to these short excerpts which are not fully explained or set out. They are merely provided here to shape a certain pattern of expectations. It is the breadth of the entire book which serves this sole purpose—not only in clarifying the dauntingly technical negotiation of the paradigm shift at stake, but also, more pressingly, to identify principles fit for at least hypothetical adoption in the transition from one to the other.[128] For, in these early texts like the Leibniz book and many of the articles in the Hermes books, Serres clearly speaks to an audience that is very well versed in both the sciences and mathematics. He travels fast, as he himself says, and sets indexes where to go in depth and look further for oneself, in order to follow and evaluate the flight of his thoughts.

Let me draw together the threads we have begun to spin thus far, by proceeding backwards toward our beginning. We set out to find an objective sense of intuition by becoming literate in a communication-theoretic reading of quantum physics (what I have also called the incandescent Paraclete).

From our preliminary discussion of Monod's book, we can extract the following ideas:

1. What we can capture objectively is not the norm but an exception to the rule, which we understand is counterintuitive.
2. The objectivity at stake is relative to the aleatory play of pure multiplicity: a code pool relative to a structure thought of as a group (not as a system).
3. Despite this counterintuitive character to phenomena, we are committed to holding on to the fact that there is something like order, however unlikely it is declared to be.
4. This is no embracing of mysticism, because such order, as an arbitrarily encrypted package of terms based upon no hypothesis in particular, is still verifiable in every aspect—even though it remains, ultimately, chance-bound.

Let us bracket the concerns this must trigger as to what "verifiable" might mean in such a situation, and for now go back to Brillouin's theorem of the Price of Information. From here we can extract the following ideas:

(1.) On the quantum level of physics, observation is not gratuitous, it comes at a cost.
(2.) The cost of observation at stake does not concern subjective interpretation of the measurement, but the objectivity of the measurement itself. In physics at quantum scales, there is no dualist separation between phenomena and the act of observation.
(3.) The cost at which new information can be acquired in quantum physics is rigorously quantifiable by a coefficient of translation between energy and information (called the Price of Information).
(4.) This coefficient suggests a law of diminishing returns: the smaller the scales of our measuring experiments, the smaller the gain of information and the greater the costs in terms of energy spent to carry out the experiment.
(5.) While these conditions subject each and every proposed experiment to judicious financial evaluation over "utility" or "demand," the more serious challenge lies rather with epistemology, the theory of how we justify what is to count as knowledge, as against speculation or opinion.

(6.) This epistemological aspect, with which Serres engages, concerns the role of chance with regard to necessity. We need to think of *necessity as originating in chance*, Serres maintains with Leibniz, and with Monod.

At this point we can ask how all these threads are woven, for Serres, into the fabric of a genuine philosophy of communication? Perhaps the most suggestive place in Serres's oeuvre to frame an answer is itself a question with which he ends the introductory paragraph to his book on Leibniz. He writes: "The system is optimistic; could we reconstruct one as beautiful, amidst the noise and the fury?"[129] With this question, Serres articulates his quest for a novel humanism—a natural, and universal, humaneness in the inchoate terms of hominescence, a *becoming* human. Incandescence is for Serres another register for the theme of freedom in philosophy—a materialist register, where freedom is endowed, "with color, with fire, and with the concrete reference to a concrete person, alive: the incandescent, or the hominescent."[130] It is a universal humanism in which human nature features only as invented persons, hence the definite article "the" before "incandescent" and "hominescent." We can think of such a notion of person as "spectral characters," but this is not the place to elaborate on this.[131] Serres assumes for his materialism of freedom, that the universe itself is inchoative, and in this sense "escent":

The Universe … is not white. It is of any colour. It is more black than white, even. But it is burning, there is no doubt about this. It originates in a primitive fire. What is interesting in the concept of incandescence is precisely that it can be applied just as well to the Big Bang, fifteen billion years ago, as to an atomic or subatomic event that will unfold in 10^{-15} seconds. It is truly a concept that permits to touch upon all dimensions of time.[132]

Such a materialist philosophy of freedom affords the development of a mathematical dictionary of how to grasp, objectively, things that remain ultimately mysterious—but this without contradicting science and knowledge, because the objectivity of such "grasping" relates to how an experience is "en/decoded."

In Serres's philosophy of communication, every object is to be considered, in a certain sense, a mystery—but there is no longer "a sphinx, seated pensive at the crossroads and from whom I must, alone, snatch its knowledge by the patience and agility of my own presumptions and the pertinence and acuity of my hypotheses."[133] The problem in science today is a reciprocal, intersubjective interrogation of "this subtle demon" that is at work in thinking scientifically, in this objective agency of thought that affects things in turn objectively, and that "never cheats." This agency, for Serres, is the world itself: "it [the problem in science] is no longer in front of me, it envelopes us and we inhabit it, we are immersed in it or it is contained by the complex figure of the encyclopedic labyrinth."[134] Words in the mathematical dictionary that index a materialist philosophy of freedom must provide a mobility in time, and in familiarity, not only in space—hence for Serres, the importance of Monod's *substantial* notion of identity as invariance that organizes in many scales, many *niveaux* of negentropy. If to furnish a word means to halt the progress of an idea, we must think of the *stasis* the words in this mathematical dictionary articulate as the precarious stasis of *mobile states*. Words in such a lexicon are *statues*. They are words that do not speak—it is the world that speaks, physically, while the words in Serres's mathematical dictionary lend themselves for *sounding how* the world speaks.[135]

Before attending more closely to the literacy required by Serres's mathematical dictionaries more strictly, it will suffice here to cite a longer passage from Serres with regard to this inchoative theory of a universal nature, this materialist philosophy of freedom, where freedom is conditioned by the objective measurement of an object's quantifiable (not its qualifiable!) "strangeness." Native to the universe, all beings count, matter, are of importance (as far as scientific knowledge is concerned) *as objects among objects*, which is what Serres's natural philosophy of communication amounts to.

> *The most revolutionary event in human history, and perhaps in the history of hominids, was, I believe, less the accession to the abstract or to generality in and through language than being uprooted from all the relations that we maintain in the family, the group and the like, and which concern only us and them, leading to an accord, perhaps unclear, but sudden and specific, to something external to this whole. Before this event, there was only a network of relationships into which we were plunged without appeal. And, suddenly, a thing, something appeared, outside of the network.* The messages exchanged no longer said: I, you, he, we, and so on, but *this, here. Ecce.* Here is the thing itself.
>
> As far as we know, those animals that are close to us, let us say mammals, communicate among themselves by repeating the network of their relationships in a stereotypic fashion. The animal indicates to or makes itself known to the animal: I am dominant over you and I give to you; I am dominated by you, so I take from you. What? It is not important or else is implicit in the relation. You are large and strong, I beg from you. Lucretius says this *of our relation to the gods.*
>
> Hence this necessitating condition, obliging animals to resolve all the problems that stem from these relations within the network. There is nothing but contracts that is their destiny.
>
> Now the human message, while often repeating the network of relations that men maintain among themselves, to the point of stereotype, says, in addition, sometimes, something on the subject of things. If it does not, it is immediately brought back to the ways and means of the exclusively political animal that is to the animal as such. Hominisation consists in this message: here is some bread, whoever I am, whoever you are. *Hoc est*, this is, in the neuter. Neuter for gender, neuter for war. Paradoxically there are no men, there are no human groups, until after the appearance of the object as such. The object as object, quasi-independent from us and quasi-invariant down through the variation of our relations, separates man from the mammals. *The political animal, he who subordinates every object to the relations between subjects, is just a mammal among others, a wolf, for example, a wolf among wolves. In pure politics, Hobbes's remark that man is a wolf for man is not a metaphor, it is an exact indication of the regression to the state preceding the emergence of the object.*

The origin of the theatre, comedy, tragedy, where only human relations exist and where there is no object as such, is as old as the origin of political relations: it plunges into animality. Politics and theatre are nothing but mammals.[136]

Quantum Literacy: Toward a Novel Theory of the Subject

The key task of a quantum literacy is how to think about the object without subordinating it to the relations between subjects. Is this not what science has been bringing us, *without* a quantum literacy, for many decades throughout the twentieth century?

Epistemology, with its three main and critical tasks of description, normalization and reasoning (foundation) of knowledge, is in the process of "losing its problematic field to the technics of science."[137] Science, as a technics, is no longer, as mechanics once was, a technics that knows how to trick natural processes (mechanics was an art, prior to the formulation of natural laws in physics with Galileo, Newton, etc.). The technics of science is one that knows how to trick scientific processes—it is the technics that leads us to replace, in the twentieth century, the idea of natural science with one we call techno-science. Epistemology has undergone a process of historicization: it turned "regional" as Serres puts it. It "explodes and distributes itself into partial descriptions, which apply to fields more and more narrow and distinct." It turned "impressionistic"; while it "describes the respective fields with greater and greater precision, its gnoseological undertaking is being put off until later."[138] In short, epistemology "constitutes itself in the natural history of science."[139] All the while, scientific technics begins to lay claims to the field of epistemology:

Modernity begins in the moment when both strings [epistemology and mathematics] inflect each other noticeably. This competition has, as we know, two reasons: Epistemology fails on the field of its former victories, and mathematics develops a taste for celebrating triumphs on the terrain of its former defeats. … The problems have remained the same [those of description, reasonability and systematic normalization], but they have acquired technical character, they are formalized and stripped of their reflexive aura.[140]

The paradigm of operativity has introduced an analogon—no, many analogons—to the reflexivity paradigm in epistemology. And this is not the end, because, "no problem can ever be solved for good. There is an essential 'historicism' in mathematics, which makes mathematics as much a movement as a system."[141] This latter point indicates Serres's relation to the critique of "scientific technics" that triggered Hegel's stigmatization of calculation as void of all thought, and prompted Husserl's interest in developing a systematic idea of cascaded retentions. It is true that forms that are operational are not only forms that are *cut out* (from their roots in truth) but also are forms that have *cast off*, as Serres puts it elsewhere.[142] The issue that Hegel and Husserl take with mathematics is exactly this *eradication of the subject position* that had been provided by language, which they see realized through mathematics with its total focus on objectivity. Serres is sensitive to such a process as well; it is true, he maintains: "A substitution takes place in which science eradicates language—this explains our time."[143] But we have been mistaken to think that language is immaterial, subjective, while only the forms were supposed to be material, objective. Language is material, and we ought to think of it as energetic: "La langue est une puissance. ... elle n'est pas, elle peut."[144] His natural philosophy of communication can be understood as addressing a novel subject position in the language of the silent words in Serres's mathematical dictionary—the forms of these words are sculptural articulations of "statements" in the statuesque terms of "mobile states":

> A reduced model of history, the rapid journey summarizes its essential gestures. That said and soon done, here's the ghost, materialized beyond the linguistic or forged collective envelope, an inaccessible stable object in the icy emptiness that excludes all life. The phenomenon or the object appears absolutely outside, as though the set of subjects had never had to deal with it.[145]

But so considered, the silent words in the mathematical lexicon index their object in a way that does not exclude life from their reference. They refer to their object as if pronouncing "here lies":

> How did the object come to mankind? In what form? The first foundation, that of the collectivity, puts the subject in relation with death. The second foundation, about which we don't know whether it precedes or follows the first, ensues from it or deepens it, puts death in relation with the object. The one makes the visible and legible face be seen, since languages vie with one another to describe it, the other makes be seen the illegible and silent face, invisible, of a founding authority that has no name in any language and that assembles the authorities that are cut out under the three names of object, death, and subject. This fundamental layer unites what lies below, what "here lies" and what lies in front. Objectivizing the subject, death gives the object to it on condition that the subject shapes it. What is this layer, this stable authority, to be called if not a statue? An inert block set there, silent, tumulary, funerary, crudely, or exquisitely worked, sometimes taking the form of a body, produced by us, exterior to us ... that stands without precession at the bottom of every origin, origins much sought-after in voyages or excavations. A first statue, silent, conditional, objective, subjective, mortuary, cast in the depths of oblivion and which bursts forth toward the Moon on the trajectory of science and technology.[146]

Forms bear nuclei of dark meaning that are not the meaning of a particular culture but that of culture in general: at the center of power, the master and subject coincide. Sacrifice is constitutive for communities in particular, the history of religion tells us, and the substitute symbols in mathematical language bear witness to this at the level of all cultures, universally: in such a literacy, with the formalities and technicalities of such a language, if we bear steady witness, sense and truth can be reconciled today, Serres maintains (this will be developed in Chapter 7, under subheading "Classicism, Remembering Contemporaneity"). We must relate the words of the mathematical dictionary to the Dionysian contents of myth, to the stories of first beginnings. Mathematics must listen to how the world speaks through the mute symbols in its language. All the while, the language of mathematics, with its puissance, is capable of a violence that must, perhaps, be called by the double attribute of rational/sacred. It is capable of silencing the world if its only speech is "in tongues"; particular, regional epistemological practices which are refused the courtesy of translation into the global lingua franca. But at the same time, it is also capable of bearing witness, with greater and lesser subtlety and sophistication, to this natural history. There is a temptation to speak of the Natural Testament with regard to the literacy at stake here; instead, I prefer to adopt Serres's term "Paraclete"—this evokes a more active register, more appropriate for the inventive-sophistication-driven, economic scope of Serres's approach with its universal domesticity, the white house (*La maison blanche*[147]), as opposed to a more legalistic, administrative scope that would validate itself through reference to a continued legacy. In any case, the universal domesticity valorized by Serres is capable of addressing the powers of abstraction at its various levels, by construing mathematical language as a language that *can*, as a language that is a puissance. Let us try to look more closely at how we can imagine this.

"La Langue est une Puissance"

"Puissance" as a term is hardly translatable in all its connotations; it renders a peculiar mixture of what in English would be called power, potency, also violence or force.[148] The thrust of a recent lecture given by Serres, entitled "La science et la langue," which is the main focus of the coming ?paragraphs, is a negotiation of the peculiar relation in which science and language are caught up.[149] Perhaps the most striking situation that Serres depicts, in order to illustrate this relation, is the following: when he began his work as a sailor, Serres recounts, a large part of his apprenticeship consisted in learning the language used by sailors. For example, he learned, to his great surprise, that rope is not part of the vocabulary in which sailors talk to each other; there is no such thing as "a rope" on a boat: "le mot 'corde' n'est pas utilisé sur un bateau. Là, il y a beaucoup de mots pour des trucs distinctes, et chacun est monovalent."[150] There are numerous words for different ropes, each one unambiguously referring to a very specific one that can be used in a very specific situation. Serres recounts that, when in his early days he could not understand the order he was given because he did not know a word used in it, his colleagues were not able to gloss that one word without recourse to an entire vocabulary of words which likewise, Serres had never encountered before. This is because of the very precise and specific (context-bound) meaning of the words that make up the vocabulary of this particular *métier* (domain of practice), a precise and specific meaning that is, in each case, embodied in the object it names.

Such words are what Serres calls "monovalent," and they are not, per se, translatable. Rope, in contrast, is a word that Serres calls "polyvalent." It is the kind of word that, he observes, makes sailors stop listening to someone who talks about the sea—because it indicates to them that the person speaking has no lived experience of it. "Rope" is an abstract word, and any explanation of a particular practice that falls back on it necessarily appears flat and lifeless, incapable of capturing any real experience.

With this anecdote, Serres introduces the distinction between what he calls *idiolects* (living languages pulsating in a particular *métier*) versus *abstract languages* ("rope"). It is worth pointing out, perhaps, that Serres's terminology differs from the usual terminology of linguistics, wherein an idiolect is the language employed by an individual. For Serres, processes of individuation cannot be separated from linguistic and communal practices. The linguist's "individual speaker" would, for Serres, be an individuating group that speaks among itself—like the sailors. There is language that, compared to particular idiolects, may be so abstract as to be almost bare of any immediate experience; statuesque language which is a precondition for the very possibility of translation, because of this very detachment. The point Serres wants to urge here is that both are crucial for the kind of repertoire we call *amphibolic practice*: practices that can be at home in several *métiers* (domains or life forms that accommodate them) at once. Such fluid communicational capacities constitute a form of living. Amphiboly in language is a term introduced by Kant as an antinomy, as something his notion of critical reasoning is supposed to circumvent.[151] Here Serres introduces into this Kantian problem a kind of historical (in the sense of generational) cyclicality: abstract words are like statues. They afford a body of memory, connecting to the origins of a certain lively language, the living moment of an idiolect. They index what preceded it and where it is about to reproduce itself anew. Such words index the mysterious darkness of a lively *métier*; they foreshadow the death that complements such liveliness. In negotiating the peculiar relation between language and science, via the notion of language as a puissance, Serres is far from claiming that idiolects are fully endowed with this puissance, while abstract language is impotent and void of it, in a sense corrupt and incapable of articulating anything that "really matters." He is also far from claiming that idiolects somehow organize this puissance of language in any original yet primitive and uncivilized fashion, as bare of sophistication, refinement and, hence, *nobility*—which I use here, following Serres, in the comprehensive sense of "being well versed in knowing" and "distinguished" (from *gnobilis*, literally "knowledgeable," from *gnoscere*, for "to come to know").

Serres's interest is not in claiming superiority of idiolects over the abstract language that propels generalization and transfer in science, as the formalization and translation of local practices. Nor is it the other way around. Serres proposes to leave behind the representationalist understanding of language in favor of a conception of language as a puissance. There is a sense here that he emphasizes, which is close to Spinoza's philosophy of affectivity: no one can know what a body is capable of, Spinoza maintained. And so it is for Serres with regard to language: no one can know what a language is capable of. And just as Spinoza elaborates his ethics through the geometrical method, so also for Serres: it is geometry that allows for locating a word's proper place within a comprehensive whole. But such wholeness is not *represented* by geometry, Serres maintains. It is actualized in the form of statuesque encryption—and it can be actualized, cryptically, statuesquely, in a great variety of ways and manners. If stasis is manifest in statues, it is nevertheless to be thought of as a particular case of movement. The stasis of statues "state" the relative symmetry of both, derivatives and an integrals; because both directions remain relative to an unfathomable origin.

The duality that Serres establishes between idiolects (language-incorporating practices) and abstract language (language providing for generalization and translation) is not an imbalance that could be stated (rebalanced) in representation. Nor does it resolve into a dialectically progressing dynamics; rather, Serres resists framing this duality and instead dissolves it into discretized clouds of possibility, multiplying and proliferating mathematically encrypted instances, each serving as a way to refer to this mismatching "equivalence-to-be." In the case of our example, that of ropes (*corde*) in the language of sailors, it is clear that topology opens up a genuine translation space that is well capable of enriching both sides—that of the idiolects (for it can introduce newly qualified manners (patterns) of knot-making into the practice of sailing) as well as that of abstract language (by rendering the novel qualifications (patterns) communicable). We can now see more clearly how Serres can affirm, with little sense of foreboding, that science eradicates language: it is not killing the vividness of language; rather, it is introducing a dimension of death into its vividness. With such eradication, language turns historical, etymological, in an evolutionary and generational sense. The proliferation of the *means* and *manners of referencing form* is what Serres understands by the eradication of language by science. It is, hence, an eradication not by disconnecting the roots from its origin, but by proliferating the roots that can connect with a word's originality. It is in this sense that he says, "A language is not, it can [*une langue n'est pas, elle peut*]." In that it needs not exclude death, silence, from what it is capable of indexing in its mathematical words, language "'can always say all' [*elle*] *est une langue parce'qu'elle peut toujours dire tout*]."[152] Thinking with mathematical words, silent ones, language is, on the one hand, eradicated, uprooted, by science. But, on the other hand, it becomes graspable how science invents objects whose roots are truly universal, in a natural, a physical and quantifiable sense. What Serres expresses here is an idea of order that seeks to accommodate both reversibility as well as irreversibility within an architectonic of thought for which amphibolic terms provide the relays for circulation, the vascular system of language's lifeblood. Such a notion of order corresponds to the nature of his communicational physics, which we will attend to in the next chapter by attempting to grasp more of this idea's scope, that of a mathematical lexicon of silent words, of statuesque words. The key term here is what Serres calls "Chronopedia." The chronopedia would be a more adequate form of organizing knowledge than the encyclopedia, Serres maintains, because it can better accommodate our treatment, and our epistemic practices, of *data*.

3

Chronopedia I: Counting Time

Meteora: The Wisdom of the Weather

The weather [le temps], current or forecast [qu'il fait], sums that which runs from the cause to the effect, plus that of probabilities, then others, distinguishable as if spread out before us, or as bifurcations, linear and circular; it combines the times of Newton, of Botlzmann, of Bergson—deterministic, entropic and statistical or bearers of improbable novelty—More, perhaps, than that of chaos. Is the weather [le temps] the summation of many measurable weathers [temps]?

Michel Serres, *Atlas*[1]

Code: A Rosetta Stone, a Double Staircase

For a few decades, science has been capable of *dating* its objects. This reveals a novel relationship to time. In what space, with respect to what subject matter, with what possibility for critique can this novel relation to time manifest itself?

Michel Serres's philosophy can perhaps be read overall as a response to these questions. He maintains that we have to think about *a science of history* that is at the same a *history of science*. This is due to a relation he calls "equipollence"; according to it, science and history are to be regarded as equals in force, power, and validity. It is within such a relation of equipollence that time begins to play a materialist role: time passes not only "formally" (linearly or circularly) but, as Serres calls it, "percolatively," "massively." It is the time that passes as mixtures percolate, as we will see shortly. Such a notion of "massively passing time" determines the "at the same time" nature of this equipollence relation. "At the same time" becomes the locus in quo where an objective relation to time manifests itself, "objective" in the sense of formal and immemorial. In order to address this locus, as we shall see in this chapter, one has to start out from and return to the strictly contemporary. *Contemporary*, meaning all that is temporary, along with all the relations it bears temporarily *with* everything else that is temporary. *With* means across the myriad scales through which energy circulates. Contemporaneity in terms of a nascent universe, whose quantum physical categories are given in the indefinite terms of maximal and minimal relations: energy's smallness (software, codes) and its largeness (hardware, massiveness). For Serres, this complex *scalarity* of contemporaneity is the domain where necessity originates in contingency, whereby the former is born from the latter—order is not the norm, nor is it to be expected; rather, it is the exception to the rule. Necessity and contingency coextend, but not impartially so: the very fact that this coextending occurs without coinciding testifies to the fact of the universe, and to that of the real world. Immanent to such coextension without coincidence, contemporaneity wells up and unfolds through staircases of scales that do not fail to provide each path of ascension with a corresponding downward path of descension. It is hence an escalating principle of scalarity that proceeds by returning to its very point of origination, in myriad routes and junctures. The scalarity of such coextension without coincidence is composed of innumerable countercurrent scales.

Serres calls the organization of knowledge along such lines a *Chronopedia*: We have to leave behind the image of time as a progressing line, and return to a placement of knowledge in a circle, he argues. But we must think of the circle itself as being vibrant, circuitous, energized, as being at once *of and in* time: such rationality proceeds spirally, inversively, inchoatively. Serres calls it *Exo-Darwinism*. Every one of the scales that balances the vibrant circle, every one of the ladders of differentiation is coupled with an inverse, and complementary, ladder of dedifferentiation. That is what constitutes the organization of knowledge within such a scenario. We will see how this amounts to the following: the lexicon for a science of history that is also a history of science (the lexicon of mathematical concepts) contains concepts that result, in each case, partially from a form of reason that Serres calls *automatic* in a sense similar to what in antiquity was called "the wisdom of the weather."

Let us begin by seeing how we can think about time with Serres.

Time Modelled as Contemporaneity

The French word for time, *temps*, derives from two contrasting roots, and has long caused disputes among grammarians seeking to ascribe a determination either from meteorology, or from chronology. As Serres elaborates in *Geometry* (1980), *temps* has the root of *temno*, for "to cut," plausibly coming from the designations of measurements and dating. But it also carries the root of *teino*, which means "to stretch." We feel perhaps more familiar with the first one of these senses, that of cutting, but the latter one of stretching has an intuitive simplicity as well, for it connotes continuous flow without tear. Behind this conflict regarding etymological roots, Serres uncovers a third domain of kinship relations, more archaic than the two other etymons, which testifies to an almost forgotten community between *time* and *weather*. This domain of kinship relations is capable of accommodating both roots:

> For *tempering, temperance, temperament, tempest, intemperate* weather [*intempérie*], *temperature*, all terms from the same family, together designate a mixture whose idea precedes, associates, and federates the two meanings— *chronological* and *meteorological*—of the word *temps*, single in the Latin languages, and corresponding to two separate terms in the Germanic languages: *time* or *zeit* and *weather* or *wetter*, languages that have forgotten or willingly left this strong community.[2]

The gesture Serres adopts here in his argument is a typical one for him: behind contradictions, dichotomies, he is looking for a domain more archaic than any etymons in particular, hence an impure and mixed origin, a matter of possibility, not of certainty. In the present case, he finds the family of words that join chronology with meteorology. How can such a mixture of two etymological domains carry any philosophical weight? The loss of this mixed genealogy is what has caused us to forget, Serres maintains—or to leave willingly behind—an understanding of "originality" in terms of a certain *likeliness*. And indeed, does it not feel much more readily acceptable that meteorology's meaning of *temps* amounts to mixtures rather than those meanings attached to chronology? Is not chronology the most unambiguous thing imaginable? Is there anything more certain than the ticktock ticktock of a clock that counts time? Nevertheless, let us try to embrace the obstacle: What are we really saying, Serres asks, when we claim that time flows (*coule*)? "With this verb we describe a flux or a river whose fluid descends, from the source to the mouth, by a channel called, just as heedlessly, a *couloir* because we want flowing to follow a channelization."[3] In the absence of any significant occurrence, a lot of water has flowed under the bridge, by which we mean the steady passage of time. When we think objectively about time, we want it to be a channel of neutral facilitation. We *figure* it, as Serres puts it, "rather like the Seine, well-behaved and rational, cultivated for millennia, docilely descending between the smooth steep-siddeness of its banks as it flows under the Mirabeau Bridge: let the night come, let the hour ring, the days pass on, I remain."[4] But language has more memory than this certain longing for steadiness, he continues. In fact, *certain things flow, while others do not.* Certain things pass through the *couloir*, while others do not: "The unity of the time that passes must be doubled into this advancing course and this immobility frozen by some obstacle stopping the progress."[5] If things do not flow, we speak of *passing* instead of flowing, and we think of the *couloir* more as a sieve than as a facilitating channel:

> For from what source does this verb *couler* descend? The Latin *colare* in no way describes the laminar descent that would bring from Charenton, after the confluence of the Marne, to and under the Mirabeau Bridge, all the water of Paris to Rouen and the Channel, but a more complicated process of passage by sieve or of filtering by strainer: on scorching summer evenings the ancient Romans used to chill their wine by making it pass through a column of snow, a term that we ought to translate precisely with "couloir," since—oh, peasant childhood again!— this very word, in French, formerly designated the funnel with a bottom of woven cloth through which we would filter the freshly milked milk: cheesecloth strewn with obstacles, not a channel of facilitation.[6]

In fact, there are always countercurrents impelling part of the flow to head back upstream, there are eddies and turbulences seizing another part under the bridge pier, randomly and in a circle, there is evaporation that transforms yet another part into vapor: "Certain elements pass while others go back up or are retained, and others lastly are annulled."[7] The verb "to flow" has reduced the physical, complex and original process to a particular case:

What we took to be the common and reasonable current amounts to a rarity. Under the Mirabeau Bridge the Seine flows exceptionally .... Here then plainly are the words: yes, the *temps* of the tempered intemperate weather or of temperature flows, that is to say, passes, traverses, sifts; so when all is said and done, time flows is translated term for term: mixtures percolate.[8]

If the water sometimes remains stable and other times heads back to the way it came from, does the formation of a pocket of memory follow the arc? Percolating time is *real* time. It is real in the sense that at each moment a multiplicity of relations can attach, or not, a large number of objects or states of affairs to each other. "Like these rivers, the world and life percolate, and no doubt, our soul lastly, and history as well, whose course is now being wonderfully drawn: an inaccessibly large multiplicity of elements maintain relations or not to each other."[9]

What Serres depicts with these images is a model that captures the activity our languages called "nature": the activity that is the aging of the universe, its *giving birth* to *time that passes*. As his etymological discussions demonstrate, the passing of time is, for Serres, always to be considered in the triple terms of time, space, and mass. Mixtures percolate because time passes massively. This is a model of nature for which mass ranges as a category among those of time and space.[10] It is the model of a physics that signifies, actively. It is a model that depicts how the verb "to flow" constitutes meaning *physically*.[11] It is a model for which signification is not immaterial, and for which the physical constitutes meaning *objectively*. The notion of model, as used here, is mathematical. This includes (but is not limited to) the applied sense of percolation employed in complex analysis (and statistical physics). "Percolation" is an important term for this field which *substitutes* the one-way directionality of functions in real and classical analysis, by relating a certain directionality to a noisy and uncertain background rather than a definite set of countable, combinatorial possibilities. The classic example to illustrate this is the following: picture some liquid poured over a porous material—mathematical percolation helps anticipate the path of absorption.[12] It is a mathematical model in the sense of being a model that is meant as an auxiliary structure, helpful for demonstrating the repetitiveness of something. It is not intended to just represent passively. It is a mathematical model in Serres's epistemological sense, where models "realize" what concepts as statues "index": like a memorial statue, the "passive" representation of statues/models is *massive*, is active. It is a model meant to (re)produce, actively through its passivity. For Serres, mathematical models in their statuesque massiveness are machinic, automatic, algorithmic. They are objective, but they are not autonomous. We will come back to this later.

It is such models that can organize, for Serres, the *dating* of the objects of science in time. All the words in Serres's Chronopedia are words that formulate silently how to date objects in time that passes massively. They are written in mathematical language, because words that can perform such a task (date objects in time) cannot be subject to definitions in the sense of "containers," but rather as "couloirs." But they can be thought of as models in this massively passive sense. Science dates its object in words that model something *massively passive*. These words articulate in that they percolate mixtures. *Time flows*, which translates to *mixtures percolate*, Serres maintains. In precisely this sense, the words of Serres's Chronopedia do not stop time in order to count it. They model the passing of time through studying how it flows—the regularity which such objects of science manifest is not that of a neutrally facilitating channel; rather, it is observable and demonstrable through study of the ways how time passes through "couloirs" (*regularly*), but "couloirs" strewn with obstacles.

In the language of the Chronopedia, science places its objects in the multiple temporal scopes of myriads of durations that "maintain relations or not" to each other.[13] They are objects bound by objective chance, as introduced in Chapter 2 (with regard to probabilistic reasoning in quantum physics). Serres grants that this feels counterintuitive. But it is consistent with his quest for an objective sense of intuition— the core of his philosophy of the transcendental objective. He maintains that after all, "intertwined, this model of the time of history ought to seem more probable and wise than the one that has us believe that history follows entirely simple and easy laws, which we would no doubt know and control by foreseeing their results, if such laws existed."[14] To commit to time as actively percolating mixtures demands a model space capable of accommodating "an inaccessibly large multiplicity of elements" that "maintain relations or not to each other."[15] Such a model space cannot aspire to represent an *exhaustive* scope of extension. But it can articulate consistent scopes, scopes that derive from coextension of what can never, actually, coincide. Ancient meteorology used to study what happens between the heavens and the earth, without having one of its orders dominate the other. This is the archaic community of chronology and meteorology that Serres excavates in his etymological gesture. Such models articulate and temper time as it passes, in an automatic (that is in actively passive) and percolating manner. Serres thinks of the model space that can offer such accommodation as *marquetry*, as inlaid mosaics made, in each case, out of time. This is how the words in what he calls mathematical language, the lexicon of which he thinks of as Chronopedia, are words whose articulations are of time and in time. The wisdom these words articulate is capable of dating the objects of science today by considering their massiveness.

> Open your eyes anew. What you see is less space than time. You see less objects disposed in a familiar understanding, rivers, rocks, mountaintops or sun, than different rhythms of a flowing, ephemeral work, secular houses, millennia old river banks, billion years old rocks, stars, millions of billions of years old. When the common representation made time disappear in space, when it dissolved or rather dissimulated, like a magician who hides a swarm of doves beneath a white blanket, when the theatrical scene of representation made it hard, for Saint Augustine, for Bergson, as well as for their successors, to envision things directly, then intuition, or the thought of duration, this series that, now, unfolds in that it lets well up, in front of me, millions of fountains, in a flash or of an infinite slowness, in front of the country house or the Grand Canyon, makes disappear, dissolves, dissimulates, in its turn, the understanding behind which appear now the chronic rhythms. ... Time has been lost in space, here now space is, in its turn, engulfed by time.[16]

Let us recapitulate: Serres proposes to attend to nature by involving a model of time that is capable of accommodating space. This is counterintuitive: we are much more accustomed to thinking of space as accommodating time. What prompts him to think of such an inverted model? Where does he anchor it? To what reality are these models of space and time being faithful? How can we think of space in terms of marquetry made of tempered complexions of time, space made out of parceled contemporaneity, that is, marquetry made *of* time *within* time? Is there not the risk in such thinking that we find ourselves carried away, by a maelstrom of massive, percolating time, plunging toward an abyss? How can we embrace so irrational a belief as that by some ladder whose steps climb up and down the universe, like the ladder of Jacob in the Bible that straddles earth and heaven? Can meteorological mechanics and chronological reasoning be thought together without subscribing to something like this biblical narrative? How can we think of models in terms of such *scalarity*?

If this thinking is to be scientific, we need to begin by interrogating how it is that measures of time can be taken in a manner that reproduces particular occasions. We must return to thinking about mechanics in non-absolute terms. Let us proceed by trying to develop a speculative picture of how time can be counted at all—with an interest that does not seek to explain the measuring of time in terms of reasonable judgement, but in terms of the cunning reason of the mechanic—a reason, in short, that is not afraid of making jumps and cutting corners. We will come back to these very important relations between mechanics (reproduction, technics) and logics (explanation, reasoning): whereas contradictions can only be held as mutually exclusive from a logical point of view, the mechanic in his craft and artifice reconciles what appears as contradictory. The terms of a contradiction (logics) appear to him as poles that need to be balanced: while the coexistence of rest and movement, stasis and motion, are to be dismissed in logical statements, such coexistence is what a mechanical object is "pulsating" with; it is what a mechanical "statement" is animated by. This, without doubt, helps to account for the ancient fascination with automatons.[17]

Counting Time: Equinox and Solstice

What is a year? It is a partition of a circle. How did the circle come to be identified as 360 degrees, meaning literally 360 steps climbed (from Lat. *degradus*, "a step [of a stair]; a step climbed")? The measure in terms of years, by which science dates its objects, takes its referent order, if we follow the wisdom of etymology-on-uncertain-grounds, from a stair that rises like a spiral. Does it lead somewhere? Back to itself, undoubtedly: the passing of years is circular. The measurement in terms of years owes itself to the partition of a circle, the geometrical figure of self-referentiality. Let us recapitulate with an eye for formulaicness: to think of a year as a stair is to think of a year as a ladder, also as a scale. Taking measure from the circle can only amount to a distribution of balances. What does it mean, then, to establish for an object of science today the temporal lapse to its origin? To "count" the temporal distance to its own beginning? *X* amount of time, measured in terms of years, means *x* times 360 steps of a stair that ascends, and at the same time leads back to its own beginning, a new cycle of itself, a new year, with every step taken.

If we think like a mechanic, we must ask now, where do the 360 degrees, which give us the geometrical form of a circle, come from? If the circle is an abstraction, what is it an abstraction from? The measurement of time must have been prior to having a scale for it. A scale also means a "series of registering marks to measure by."[18] Let us ask then, what marks are we talking about, and what are they registering? We are trying to imagine here what could count as the original referential order that disposes to a treatment of time *distributively*. It is the key thought that produces an idea of passing time without having to freeze its massively passing contemporaneity (the coextension of durations whose sum does not produce a co-insistent total of all those durations).

Those who remember what antiquity called *the meteora* will remember the centrality of equinoxes and solstices for meteorological wisdom: Latin *aequinoctium, aequus* stands for "equal," and *nox*, genitive *noctis*, for "night." *Solstice* is derived from the Latin *sol* for "sun," and *sistere* for "to stand still," because at the solstices, the sun stands still in declination. That is, the sun's path varies with the seasons, and the seasonal movement of the sun's path (as seen from earth) comes to a stop before reversing its direction. Counting how many times the sun rises helped build anticipations of the seasons. There is not one particular narrative for *whence* or *why*, twice a year, the days and nights are almost precisely of equal length, although there must have been cults in plenty that attempted to explain this. But the phenomenon itself was so plainly given to intuitive grasp—just an observation of the skies.

The Turning Points for Modelled Beginnings and Ends

How did people observe the sky in antiquity? They did so with the help of what was called the gnomon, the sundial. Serres explains:

> The needle of the sundial projects its shadows on the ground or the reading plane according to the positions of the stars and the sun over the course of the year. … The light that comes from above writes on the ground or the page a drawing whose appearance imitates or represents the places of the Universe through the intermediary of the stylus's point.[19]

This device was not in all likeliness much used to count hours, he maintains, for "in those times, hours varied, since the days of summer or winter, whatever their length might be, were invariably divided into twelve: a bad clock, this gnomon!"[20] But the gnomon served as an instrument to observe the sky, an instrument for scientific research that gave a model of the world, "which would show the length of the shadow at noon on the longest and the shortest days, from which solstices, equinox and place latitude for example were drawn: thus more an observatory than a watch."[21]

Like a ruler that gives knowledge, the gnomon etymologically signified "what understands, decides, judges, interprets or distinguishes,"[22] but not in the sense of an individual cogito, organized into generic faculties—those imagined organs in our heads that supposedly act as receptive support for all that can be learned (the mental faculties of memory, imagination, and reasoning). The gnomon is a ruler that affords knowing in the sense of a "machine" for knowledge.[23] It is *mechanical*, and hence renders social forms of understanding *facultative*—in the literal sense of facultativity, namely, that it *substitutes* them. The gnomon's knowledge is objective, not subjective. While the equinoxes and solstices are our turning points for *modeled* beginnings and ends, the *staging* of the natural shadows and lights takes place through the "interceptions" of the gnomon—*rendered* by the gnomon. Serres explains: "This point writes unaided on the marble or the sand as if the world knew itself."[24] The knowledge this object produces is mechanical knowledge in the sense that it indexes a natural movement (the course of the sun) through deployment of an obstacle, with the intent that such interception returns objective results (the formulation and formalization of latitude, solstice, equinox, tilt of the world's axis) in precise (measurable) terms—all of this *without* human intervention, rather, "like an automaton, without a motor subject."[25]

Serres sees in the gnomon "the first machinery uniting hardware to software."[26] This is because together with this mechanical device that automatically records, and which is capable of telling time geometrically, comes an artificial memory that must be called objective: with the gnomon, the observation of the sky yielded a *graphism* of the world that operationalized the orders of the skies in the service of mapping the world:

"The calculation of latitudes according to the shadow of the Sun at the solstices and equinoxes—the first mathematical link between astronomy and geography— … gave rise to the establishment, by Ptolemy or before him by Hipparchus, of what Antiquity called the tables of chords: long lists of ratios between the measurements of the sides of right triangles and the measurements of their angles, in which the birth of trigonometry can be read."[27]

Such tables of chords follow the same reasoning as Pythagoras' theory of music, and of number: they produce models by help of instruments. In the case of Hipparchus and Ptolemy, such instruments were the gnomonic observatories, while for Pythagoras it was the monochord. We will come back to this later.

Of Tables and Models

It is the models, in their objectivity and in their capacities, that are *wise* in this thinking; in the accounting they facilitate, they connect automatic, algorithmic reason with the tables of records that grant mnemotechnics. Like the wisdom of circular reason in terms of scales, which takes the equinoxes and solstices—the abstract points of equal duration, and that of reversal in direction—as turning points for modeled beginnings and ends, the wisdom of such models too is not solely nourished by explanation alone. This wisdom does not seek to calm worried reason by flying daringly toward an *ultimate*—a *first* beginning, or a *last* ending. It is a wisdom that is circular, and that nevertheless grows more mature, and more powerful, from paying close attention to phenomena in their varied yet iterative returnings. It is a wisdom that gives birth to the ordinary *as the genuine rarity* which every ordinary thing is. Serres's thinking about the gnomon is profoundly informed by his notion of time in which chronology and meteorology are wedded. Time passively percolates mixtures, if we take mass into account among the basic categories of time and space. Time stretches continuously, *and* it is being cut up. The wisdom of models at stake here is the wisdom of an automatic reason coupled with mnemotechnics that depends upon no subjective cogito. It is a wisdom that is objective, automatic, and yet this wisdom is not autonomous—it depends not upon a subjective cogito but upon an instrument that can be played more or less well.

Such a notion of instrumentality is crucial for imagining an objective reason that depends upon an objective sense of intuition. Such instrumentally augmented, objective reason thinks by incepting a covering space.[28] We can picture Serres's speaking of a marquetry of temporalities, his space of the Chronopedia that is made up of inlays referring piece by piece to each other, in terms of just such covering spaces. The space of the Chronopedia constitutes the locus in which scientifically datable objects can be placed. Objects are addressed in this locus by words that are capable of modulating and articulating the reproduction of certain events mechanically. They are the words of mathematical language, a language that exists not only by extension in time and in space but also massively. It is a language that is a puissance. Its words are statuesque; they introduce death into the vividness of communal habit. Such statuesque words perform like instruments that can be played in a manner that opens up an indefinite qualitative scope of effects: statuesque words of mathematical language are abstract words. They can do something without defining it. They are performative, but not autonomous—like musical instruments. The locus organized by the Chronopedia is the locus where contingency accommodates necessity, whose categorical order (its lawfulness, the addressability of locations in such a locus) is evolving gradually— orchestrated across many scales. It is more or less stable, more or less restless, here different from there, varied everywhere, overall kept up by a wisdom that is as diligent and artistic as it is automatic and mechanical. To model the passing of time in this way, massively, is bound also to bring to light instances of phenomena that are *not* likely to occur. It is a model that represents only if it models its object *and* the object's inverse. This is what makes it a mechanical model—in the sense, as I will argue later, that ought to be called *architectonic* (Chapter 8, "Coda: Architecture in the Meteora")

The circular wisdom of models of time's percolative passing is a wisdom that draws direction from returns that make sense to it, and that gives sense back to unlikely returns—returns that it, nevertheless, appears to have drawn from itself. From the point of view of the mechanic, all instruments that act as scales are such models. Scales are drawn from partitioning the circle. Scales are fashioned to bridge the equatorial turning points in modeled beginnings and endings. The year, as a scale with approximately 360 steps, beginning with itself and ending too with itself (its *aged* self), can now be imagined as a model in the sense developed earlier; a model that renders asymmetrical relations balanceable, "temperable," like a musical instrument does. So considered, scalarity is what intercepts—indexes, codes—a cycle of reciprocity. Scalarity is therefore also what makes this cycle bifurcate, what renders it circuitous, what makes time pass in massive percolation: counting time in terms of years allows us to date an object by "stating" its age scientifically.

Let us consolidate what we know so far: countable time hinges on two equatorial points. Time is countable through scales that partition the equatorial circle. No narrative explaining the mysteriousness of these points is necessary for measuring and counting time objectively, mechanically. Scalarity, if we understand it as being born out of an inherent association with those mysterious (yet objectively registered) equatorial points in time (equinoxes), together with those singular points at which passing time reverses its course (solstice), introduces a notion of model that does not represent an order in general; it realizes an order, situationally. As such it is a model capable of generalizing the augmentation of a local constellation of relation (think: any artifact, for example, a violin, but also a watch, a luster, a table as embodying the generalization of such augmentation). It is the notion of a model for which artifact and object are not categorically different things: an artifact in mechanical terms is independent of subjective will, but not of cunning thought. It is entirely objective, and yet there is more or less sophisticated reasoning/performance to it. At the core of Serres's philosophy of the transcendental objective is the question of how such sophistication in reasoning can be addressed. We will see a bit later how Serres's interest in structures is motivated by the same question; structures to him are models that are, in this scalar sense, discussed here, "formal through and through."[29] They generalize (and mobilize) not an order, but the augmentability of a constellation. They are models whose formality brings us in touch with "more" of the *contingency* of the real world, they are not estranging us from it. They are models that depart from the real world's "dark meanings"[30] without having to sacrifice reason and rationality in favor of fate or irrationality. Relating math and myth means dealing with scalar models that are reliably capable of learning from singularities. Indeed, as we have seen in this chapter, scalar models actually owe their very existence to singularities (the singularities of equinoxes and the solstice). With regard to equatorial scales, we can straightforwardly say that the rarest and most unlikely thing gives rise to the most stable notion of order and repetitiveness.

Sense Means Significance and Direction

The conception of counting time that Serres advances with his discussion of the gnomon insists that orientation comes from *scales*, not from an *arché* or a telos. Scales need be referred neither to a past nor to a future. They afford the keeping of records of *the advent* of things that make sense, from Latin *adventus* "a coming, approach, arrival to, arrivals." The records of scalar data tables behave like sieves (or spectra). They facilitate the percolative passing of massive time. Counting time *makes* sense, yes, and sense here literally means "to find one's way," or "to go mentally."[31]

Counting time offers a sense of scale that is *not* drawn from the empirical world alone. The data achieved by counting time does not represent the empirical world without, at the same time, affecting it. Scales so considered draw direction from themselves and also give direction to themselves. They propagate a wisdom with which the circular, through its partitions (the scales), sources a measure by unfolding it from itself. Intercepted by scalarity, the indefinite cycle of reciprocity characteristic of passing time *accommodates itself* in the locus in quo referenced by the Chronopedia. The indefinite cycle of reciprocity partitions its own magnitude in the given phenomena of the world. How can we imagine what this means? Let us make up an example to illustrate the gesture even if, for the sake of simplicity and the suspension of technical precision, it is somewhat frivolous. We can say that in the wisdom encapsulated by the Chronopedia, passing time derives measures from itself to contain conductively a percolating mixture into a complexion (we can call it a formal "box" rather than a definition) that indicates, from experience, a certain likeliness for something to happen. For example, a peasant notes a certain breeze, notes together with it a particular temperature and the singing of particular birds, and he expects from experience that when it occurs, as a *boxed complexion*, it is likely to bring, for example, rain. What this means is that every measurement in terms of scalarity incorporates a certain *objective*—even if chance-bound—mastery. Of such mastership we can say that it does not *submit* the empirical world to its domination in any linear and simple way. Such mastership is mastership that knows how to welcome the real world at its tables (the mnemotechnical tables of records); the tables that collect data produced by circular, scalarly partitioned, objective wisdom.

In antiquity, such knowledge belongs to the discipline that studied the meteora: a wisdom that reasons—judges, even—with equatorial scales (equinoxes) when the sun stands still (solstice), and that accepts the self-righteousness of the scales as a mysterious, but natural, power. A power of indeterminate (because circular) but, to a certain and always imperfect extent, *recognizable* direction (sense). Objects that incorporate direction in this scalar manner make sense actively, due to the fact that the measurement of the temporality they incorporate "knows how" to percolate mixtures. They are objects that make sense because they are directed, but they are directed in a transcendental manner. They are, hence, *objectives* (due to the direction they incorporate, the direction that can be read or deciphered as the sense such objects can *make*). Serres's philosophy of the transcendental objective confronts us with a notion of object that is biased—marked by *chirality*, literally meaning three-dimensional forms not superposable on their mirror-images, like left and right hands, the two halves of a brain, and so on[32]—and hence needs to balance a discretized (symbolized) and inverted equatorial symmetry. They can do so in a great variety of ways and manners. We have thereby a teleonomy that is indeterminate but nevertheless directed. This is the aspect of Jacques Monod's natural philosophy, in *Chance and Necessity,* and from the point of view of molecular biological, which is so important for Serres's philosophy at large. Monod's key point, with regard to an evolutionary point of view, is that we need to think a lawful telos that acts from a distance in relation to a code pool, a quantitative reservoir that can be ciphered (partitioned into numerals, "chiffres") and hence counted in many ways, which he calls an invariance. The crucial point is to think of it as a *resourceful capital* rather than as *a stock of resources*. Every object formulated in these terms, the terms of such a transcendental philosophy (according to invariance and teleonomy, that is, on equatorial scales and under the assumption of inverted direction [solstice]) is what we can call a *saved* phenomenon— it is a singular event saved, from time that passes massively, through the sophisticated accounts one is capable of giving of this phenomenon's strangeness.[33]

Meteora

The meteora stood for all that happened *between* the earth and the sky. It was the domain for studying the seasons, the temperaments of weathers, from the point of view of life on earth: The *meteora* consisted of the sum of all measurable, and thus articulable, temporalities, durations, seasons.[34] The domain of *meteora* as an intelligible phenomenon is universal; there is no place on earth where it plays no role. At the same time, *meteora*, in the ramification of phenomena, is varied, local, diverse, multiplicitous in every corner of the planet. The nature of *meteora* requires measurement understood in terms of a sum, a total, but this totality knows nothing of an *ultimate*, neither first nor last. It is a total that does not homogenize the things it calibrates, but one that sums toward their universal identity through as many scalarly measured—*saved*—phenomena as it is capable of. The total of such summation is a total that grows from its own exhaustion. It is a growing total, one that sums up the wisdom of what can be known in *terms* that count as *terminals*, never as ultimate "terms." It is the wisdom of complex, mixed, heteroclite phenomena that can be brought to light by demarcating them, so as to preserve them for what they are worth, so that they can be looked for, recognized, and placed or welcomed. The following is a beautiful passage from Serres that illustrates this point:

> An old peasant scene: every morning upon waking, before deciding what work to start upon, the farmer examines or observes the sky and tries to assess, predict, evaluate, weigh the intemperate weather that awaits him, a problem that plunges his temperament, touch, sight, smell, memory, into a formidably complicated mixture of wet and dry tempered together, of cold and hot, allied, making up the temperature, of long and short, of continuous and broken, whose present promise permits avoiding the ravages of the tempest or defines—for plowing, sowing, grape harvest or hay-time—beginning or ending, that favorable moment the Greek language calls *kairos*, from a verb that again signifies mixing.[35]

Who was the owner of such knowledge? Who was the privileged subject of such wisdom? In the tradition of the *meteora*, the cognition—the discerning, the intelligence—at work was attributed to an object: the gnomon. The sundial was a stick, erected upright toward the sun. It could be placed so as to catch the shadow of a particular phenomenon's temporality (its duration, its temperament) by indicating the hour of its appearance and disappearance. The gnomon facilitates theory building that attends to the rhythms in nature, unfiltered by juridical or political theory. Nature, as far as the theory of *meteora* was concerned, was *automatic* in the literal sense of self-acting, moving on its own. The gnomon was acting more and more precisely—without thereby losing its mysteriousness. The gnomonic space of observation *nests* in time, forms pockets that *know*, physically, *how to preserve* secrets throughout the days that have passed. This idea will be developed in the next section.

What we are trying to demonstrate is that the wisdom of the gnomonic space of observation can be thought of as coextensive with the knowledge of the *meteora*: its space is the manifest spacing-out of maturing, self-referential wisdom. Such wisdom is of a double nature: it is *immediate* wisdom with regard to the objectivity that the gnomon's records provide. And it is *mediate* knowledge with regard to how those records were administrated in religious, political, economic, juridical rituals. This is how it connects, as Serres puts it, *le logiciel avec le matériel*, software with hardware, energy in small scales with energy on large scales.

A Logos Genuine to the World: "Le logiciél intra-matériel"

Software, Hardware

Expressed in any language whatever, it makes no difference; geometry writes a universal language common to "the peoples of the world who reason with it."[36] The intelligence proper to geometry is *objective* intelligence. Subjectivity emerges through the insertion of this objectivity in any one of the locally diverse, nonuniversal, actually spoken languages. From this point of view, every language is an artificial intelligence trying to live up to, to appreciate and translate, the capital stock of an objective intelligence which Serres calls "white," and of which he maintains that it is "immanent and no doubt coextensive with the Universe."[37] The hardness of the material gnomon as a stick coexists with the softness of gnomonic reasoning in a projective, and objective, space of observation. Together they constitute the kind of reason Serres attributes to the wisdom of the *meteora*, reason that he calls *automatic*. Automatic reason, we need to remember, is considered algorithmic but not autonomous by Serres: it always comes with, and depends upon, the sophistication incorporated by a particular mnemotechnics: "an algorithmic thought always shows two components, one that can be said to be mechanical and the other which must be called mnemonic."[38]

For Serres, the world, according to this model that does not represent the world immaterially (without affecting its own self-presentation), is considered a material black box; in its reality, the world is massive and dark, comprised of percolating mixtures (hence its darkness), but nevertheless also containing a certain wisdom. This is so because *in* its counted measuredness, its measured accounting, it forms a "complexion." There is form *and* darkness to a black box. The world so conceived filters itself, passes on, in its complexion-ality, some of this "white intelligence" of which Serres maintains that it is "immanent and no doubt coextensive with the Universe."[39] This white intelligence is somewhat mysterious, as it sounds in these formulations. But it can be referred to also quite concretely, as natural according to quantum physics: Serres ascribes this intelligence to the white light that is emitted by a sun. Another way of talking about such a black box is to say that the model describes *mechanically, quantum physically*, how the natural world "behaves." The "natural world" (not "nature"), let us remember, means the objects insofar as they have an age that can be *dated* scientifically—including the age attributed to the universe itself, that of the earth itself, the appearance of life on earth, and so on. To appreciate this perspective, it is crucial to understand that for Serres, information resides in *form*, not in material.[40] In Chapter 2 we developed how, for Serres, information theory after the quantic turn provides for a communicational approach to physics according to which information resides—massively so (hence a physics, not merely a formalism) in form, not in matter. He also describes such a communicational physics as an "intra-material logicial," a kind of software at work within materiality itself. In this section, we want to see what is therefore at stake. Let us listen to Serres. More radically, as he puts it, matter, as the opposite of form conceived as immaterial, does not exist at all:

> There isn't any matter in the Universe. Otherwise the physical sciences would have ended up encountering limits to their progress or their history, boundaries foreseen and placed by materialist metaphysics. This latter vanishes with the progress of the physical sciences, which never cease discovering forms without ever encountering any matter they don't name, so as to only recognize mass. Matter doesn't exist; only forms are found, like atoms, and all the way down to the tiniest particle, with or without mass, innumerable forms, as well as their chaotic or ordered mixture,

a system or noise which tosses and shakes their innumerable multiplicity as in a basket. There is only information, whose enormous stock in the world, no doubt expressible by a very large number, mathematically finite but physically infinite, leaves science in an open history. Even weight codes a field of forces, even any aggregate, colloid, or organism recodes a subset of coded forms. Only mixture and disorder, noise, chaos, give the illusion of matter.[41]

As we have seen, information in terms of information theory is massive; it is electric. This in turn, for Serres, also means that mass never exists purely, but only as "informed." Unlike matter, Serres maintains, mass does exist in the universe, but never purely—mass, different from matter, is attributed an order in terms of entropy. Mass only exists relative to a particular entropic order's negative, its negentropy. Pure mass would be the absence of information: it would be the ultimate oven, where light is, physically, a burning fire.

This distinction between matter and mass, drawn from information theory and its quantum physical terms and insisting that with regard to what we can know, mass and information only exist as mixtures, is important to understand with respect to Serres's use of the terms *hard* (materiél) and *soft* (logiciél). In *The Incandescent,* there are two clear passages on these difficult notions. The terms soft and hard actually both refer to one and the same thing: the contingency that the "initial conditions" of the universe are capable of "organizing." Serres calls this ultimate and original contingency *la circonstance.*[42] The "hard" (*le dur*) is "*circonstance* in its maximum"—it is the universe most proximate to what physicists call the Big Bang. It is ultimate and original causality. *La circonstance* would be *immediate* reason. Relative to this, Serres calls "soft" (*le doux*) the process by which this ultimate and original causality is coded in nature, in the differentiation, in the becoming and the passing of beings through *current states.*[43] Serres's *circonstance* names in spatial terms contemporaneity as the nascent space, where necessity results from contingency, and where they coextend in reality without ever actually coinciding. He explains:

> One must come back to becoming. Let there be a certain current state of a system in evolution: I make no hypothesis about its material: solid or fluid, liquid or gaseous, living or cultural, social or linguistic, virtual, mythic, artistic, whatever, subjective even. It changes, it evolves along the duration. The procedural ["*le processual,*" in German "*die Prozesshaftigkeit*"] hardly ever follows a single line, but rather unfolds in a space-time of several dimensions. It may encounter a circumstance that makes it bifurcate: a hard barrier, viscous tributaries, electric tension, heat, cold, microbe, obstacle or unexpected additives, transformation of the environment, and occasion that the Greek called Kairos, lightning strike, thunder, or love, Cleopatra's nose. The circumstance in question restructures the common state.[44]

Information results from how the world reasons—in reality, and with rationality—which amounts to how the world reasons in terms of finitude: "The world adds up and gives an enormous stock of forms. ... The entirely informed forms lie in the things themselves, where it suffices to collect them; thus our works reverse the ancient processes by which information only came from our skillful hands or expert understanding."[45] The world, in terms of this white intelligence that coextends with the universe, which Serres also calls "le logiciél infra-matériel," engenders sense. Information resides in informed mass, not in a materiality taken as the opposite of immaterial forms. Serres explains: "There exists an immense objective intelligence of which artificial and subjective intelligence constitute small subsets. Our intelligence is not an exception in black surroundings that would passively wait for us to inform them. The object that we know is forged by us in a way that's analogous to certain things of the world, forever our guides."[46]

In the nature of the world, the *logiciél* and the *matériél* consist of mixed quanta of mass and information that provide islands of order (negentropy) amid, and countercurrent to, the entropic emission of solar light (energy without mass) that animates all things. The massive passing of time tempers its light speed into the durations proper to the things of the world.

Serres thinks of the two principles, the hard and the soft, from within a framework of maxima and minima: the language of geometry is universal, and to exhaust the objectivity it is capable of expressing, he says,

> a more than adamantine hardness, infinite, and a more than aquatic, aerial or ethereal softness, infinite as well, are required for this earth whose material or special consistency causes the infinite of a maximal resistance and the infinite of a minimum of light breath to become equal in it: therefore so hard that it includes all possible hardware in the universe, therefore all the applications of physics, astronomy, chemistry, biology … so soft as well that it makes understood all the software of the universe, languages, signs, symbols, notes and musics.[47]

Economy of Maxima and Minima: An Anarchic Logos

Where is this universal language in which geometry is written *native*? Where does it come from?

What we need to do is to try to imagine an economy—circuitous, yet generational, aging, and scattering exchange in a communicational, reciprocating physics—for which no purity or original balanced state need be assumed as a reference. This can be achieved in terms of scalar models that straddle maxima and minima points, as with the diurnal mapping of the year where the two *maxima* are the days in which day and night are of equal length, and the corresponding *minima* are the days where their difference is maximal (or the other way around, it is not important whether equality or difference is considered as maximum). As for the relationship between mass and information, Serres proposes to distinguish between points of maximal hardness and minimal softness (one side of the scale), and maximal softness and minimal hardness (the inverse side of the same scale). I suggest that we can call these points "quasi-equatorial." Let us develop how this presents us with an economy that affords an *anarchic* reasoning.

Geometry produces a space, in the cyclical terms of a polar system, within which every position is a vicarious position. It is the reason of such a universal economy, in terms of maxima and minima, that produces a logos proper to the world; a logos of an import and export economy that the world maintains with itself. An economy in terms of maxima and minima can be considered a form of reasoning, since it proceeds by measurement (via scales) and since it produces models (that afford anticipation). This is one of Serres's decisive ideas, namely, that geometrical reason is reason governed by a principle that insists by definition on the vicariousness of its own position, that is, a principle that demands its own substitution: geometrical reason "demands that there be no reason."[48] This is not as paradoxical as it sounds at first: geometrical reason is the reasoning that *renders apparent the transparent structure* by which the inner logic of any principle (a principle's *principality*) governs its domain. He explains:

There is no arché except in the archaic. Science, arriving on the scene, preserves it, brings it to and imposes it on our reason, blinded. … The beginning expressed by the term "archaism" is found again in the command of the word "hierarchy." Can, conversely and in general, an anarchical system be conceived, without reference or border deprived of privileged place or referential, and yet rational?[49]

Serres's answer is straightforward: "Yes, assuredly: it suffices to trace back to the multiple variations of beginning in Anaximander's indefinite. Things begin when the arché precisely goes absent, and command appears when they claim to begin."[50] But, he maintains,

The model of the world by the same Anaximander nevertheless doesn't separate reason from the old mastery or from the archaic hierarchies. … Hierarchy remains inside reason, but since height, power or king are no longer spoken of, it becomes transparent inside reason, so invisible that no one has seen it, that no one thwarts this intelligent Greek ruse. … The hierarchy remains transparent in translucent reason; they become identical to each other; an admirable trick, power lies in knowledge, the way the invisible lies in what allows seeing.[51]

How can one think of a geometric reason with regard to an economy of maxima and minima where mixtures of mass and information, of hard(ware) and soft(ware) are at stake, where "translucent reason" becomes "addressable"? We begin with contemplating another citation by Serres, where he talks about the origin of geometry with Thales, at the foot of the Great Pyramid, the story which recounts Thales's invention of his famous theorem regarding the variability of the angularity of triangles within the scope of a circle:

By relating the tomb's shadow to the reference post or to his own shadow, Thales states the invariance of the same form across variation of size. … Thales shows the extraordinary weakness of the heaviest material or hardware ever prepared in the history of men and nations, as well as the omnipotence, in relation to the time that passes, of a certain software [logiciél]: of the logos itself on condition of redefining it, no longer as speech or saying but, in lightening it, as same relation; even softer because the terms of the analogy balance each other, because the one is erased by the other, as though each one nullified the meaning of the other, what it states, its import, so that that only their pure and simple relation remains, the common form of the statement.[52]

Thales did not like kings, it is said, and Serres reminds us of this. From the maximal remains of the maximal of optimally preserved history, Thales draws minimal softness or lightness: "A miracle, truly: from almost nil means the longest of possible empires is born, that of Mathematics, which laughs at history without henceforth knowing decline."[53] We are scarcely beginning to assess such an *economy* that remains faithful and just—in a manner we could call "civically anarchic"—to the principle of reason according to which it governs: that there be no reason, that there be no formless matter.

Let us try to make acquaintance with the logos of such a quantum-literate reasoning in the form of an exercise in concentration. Let us try to acquaint ourselves with the gestures in thinking required by such an anarchic civility by exercising ourselves in how to endow formulations with indexes that do not aim at having the last word, but at affording a certain conviviality through being not, altogether, unreasonable.

4

Chronopedia II: Treasuring Time

Homothesis as the Locus in Quo of
the Universal's Presence

*Thales, reading and recording the traces of the body, deciphers no secret other than
that of the impossibility to enter the arcanum of the solid body in which knowledge
resides, buried forever, and from which springs forth the infinite history of analytical
progress.*

Serres, "Ce que Thalès a vu."[1]

First Iteration (Acquiring a Space of Possibility)

In Serres's text,[2] we find ourselves in the desert with Thales, facing, in the pyramid,
an impenetrable constellation. We might well recognize the pyramid's outline as a
triangle, but we know not how to measure it. We are taken to accompany Thales on
an adventure that is pure concentration, a tour during which we reach, eventually,
in a circuitous manner, what is straightforwardly and directly inaccessible—a space
in which measuring the pyramid becomes possible. It is an adventure in *methodical
an-archism*, in reasoning that proceeds by an act of double duplication: on the one
hand, we duplicate the situation in which we find ourselves, and on the other hand,
simultaneously, we duplicate ourselves as we find ourselves comprehended in that
situation. All that is left for us to do, if we follow Thales and Serres, is to give an account
of how we proceed by aspiring to measure each repeating step taken. The cunning that
drives such reasoning never properly manifests itself, either positively or negatively. It
establishes on lofty ground, through what I will call *double duplication*, a market(ry)
of abstraction that is capable of hosting the play where homothesis provides the
establishment of a "homology between the crafted and the craftsman."[3] The cunning by
which we are driven manifests in no other way but in tending to its own continuation.
Tended by his own cunning, Thales's double duplication introduces a time that might
remain, by giving way to the unlikeliness of finding an accord in which *it* (measuring
what is overpowering, colossal, and immense) acquires a space of regularity from sheer
contingency, exposed by elaborating the soundness of the presumed accord through the
computation of auxiliary structures in all of that in which the same invariant quantity

is at work. The postulation before Thales's inner sight—a postulation in theory—of a module, from Latin *modus*, literally "a measure, extent, quantity, manner," is enough to stage the invariant quantity at stake as a market(ry). This is what Serres is telling us.

But how to find this quantity? All that there is to be contemplated, for finding an answer, as we remember Serres telling us, is that Thales must find a unit of procedure, and that the quantity of this unit ought to be, if the procedure is feasible and valid, conserved by a structure. Thus, Thales must attempt to *stage* abstractly the very act of virtually "en-familiarizing" himself with what is colossal and immense. Thales knows that the interiority of the pyramid is inaccessible, that it would be an unworthy violation to force his access into it. Thus, Thales pays all due respect to that, and premises for his own symbolical double duplication that the interiority spaced in it be inaccessible as well. He treats the size of his triangle purely structurally—without knowing, at first, anything about the structure or how he could possibly apply his triangle for measuring. We thus learn that Thales begins this elaboration by building a stock of experience— Serres calls it a *résumé*, from Latin *resumere*, "take again, take up again, assume again." Before Thales will be able to actually draw a circle, we learn, he has to actually go in circles. Many times. Learning to measure, even in theory, Serres maintains, is an operation of application. One has to "blossom into" the capability of doing it. Thus Thales keeps beginning, summing up what he finds along his iterations, and treats the sums he comes up with as a product of reciprocity, from *reciprocus*, "returning the same way, alternating." Gradually, we are told, he invents a scale of reproduction. How? All that we can say in this first iteration is that Thales measures the pyramid by postulating—on grounds no more "solid" than the immateriality of a desire—that it *be* possible, and by striving to elaborate the conditions for his own postulation.

Second Iteration (Learning to Speak a Language in which No One Is Native)

One idea Thales substantiates in the course of the elaboration of his postulate—that the inaccessible pyramid is measurable—is that the pyramid incorporates the principle of homothesis. Homothesis is, as we learn from Serres elsewhere, "the same way of being there, of being placed."[4] The space of homothesis is a space of dislocation, deferral, and adjournment, "with or without rotation," as he puts it.[5] Things that are governed by this principle, things that are tributary to the space of homothesis, are things that can be considered bounded by form. In short, they can be considered as things that are commensurate. But what can be the source that sheds light onto such a space for abstract intellection, and hence open it up to our intuitive sight? It is the sun that treats all things equally. Yet this equality, Serres warns us, cannot in any direct manner be found in the sun itself, as if it gave each thing its natural gloss immediately. Nevertheless, we are told, the sun facilitates the engendering of an abstract space, which, for Serres, corresponds to the Greek "miracle," whose revelation eventually made possible what he calls the fabrication of a mathematical language, the sole language "capable of halting conflicts and which never needs translating."[6] The language spoken in such abstract space is the sole language in which there are no barbarians, because everyone speaks it as an immigrant, with no political obligations of conforming to a mother tongue spoken by natives.[7]

Third Iteration (Setting the Stage for Thought to Comprehend Itself)

This language allows articulations on the stage of abstraction, and for Serres, its possible articulations open up and constitute the market(ry) scene of writing. Within a space governed by the principle of homothesis, the scene of writing is constituted around homology. For Serres, it is the Greek understanding of logos that will allow alphabetic writing to think of the cosmos no longer in terms of genesis and progeny, but in terms of a logics that comprehends the cosmos *within* the universe. Homology, he tells us, is threefold: number, relation, and invariance—arithmetic, geometry, and physics. This fantastic premise of one universal logos, Serres maintains, allows Thales to see in the pyramid a manifestation of the homothetical principle. On this assumption, Thales can postulate the *invariance of form* to complement the *variations of quantity*. Armed with such thinking, the colossality of the pyramid becomes less daunting, and this without the need to divest its constitutive secret, its inaccessible interiority. The archly reasoning that supports such thinking is not the reasoning of an individual subject rising up against the principle that governs its own predication. In Thales's circuitous thought, there is nothing revolutionary here whatsoever. The reasoning exerted in support of homology is an automatic reasoning, we are told, from *autos* "self" + *matos* "thinking, animated." As Serres puts it, it is the reasoning that happens as the world exerts itself upon itself,[8] a world that thrusts forth and pushes out of itself, in order to adjoin to itself what happens to it. Serres calls this the reasoning of how the gnomon counts, the reasoning that seeks to account for the objective ruler that sets the natural play of shadow and light in mise-en-scene by collecting it with its own apparatus of capture. The scene is objective, but it is a scene of markings, and as such, it is a scene of marquetry. "Who knows? Who understands? Never did Antiquity ask these two questions," Serres maintains.[9] The gnomon permits the indication of time, but first and foremost it is an observatory that does not, like modern telescopes, bundle something specifically for the sight of an individual subject. In the events the gnomon is capable of staging projectively, Thales (and anyone else) participates as nothing more than as a pointer, an index, or cursor, since "standing upright we also cast shadows, or as seated scribes, stylus in hand, we too leave lines."[10] But aware of this precise circumstance, Thales now sets out to reason about how the gnomon stages projectively, as an apparatus of capture, the play of shadow and light. In his double duplication, Thales literally tries to catch up with the course of what he himself (as a gnomon) indicates, and hence makes observable. It is by trying to catch up with his own significance within the situation that Thales eventually begins to substantiate the concept of similarity as an invariance—or, to make Serres's point clearer—as an idea contemplated by the world in its own automatic, accounting reasoning. Even though Thales is trying to catch up with his own significance within the situation, the active center of knowing resides outside of Thales himself: "The world renders itself visible to itself, and regards this rendering of itself: Here resides the meaning of the word *theoria*. To put it more clearly: a thing—the gnomon—intermits the world through stepping in, such that the world may read on its own surface the writing it leaves behind on itself. Recognition: a purse, or a fold."[11]

For Serres, the scene of writing is automatic. It unfolds for him on the stage of projective abstraction. It is an economic space, and this does not mean that it is not at the same time also a space coordinated by a point of origin—but one that remains mystical. But it too is a space that knows no individual poets or playwrights. The dramas it puts forth are authored by a collective subjectivity that spells out the reasoning of a world that exerts itself upon itself.

Fourth Iteration (Intelligence That Is Immanent to and Coextensive with the Universe)

Such a collective subjectivity depends upon an artificial memory, which Serres finds in the canonical lists and tabular organization of practical problems—the preparation of how certain results to certain problems may be found more easily, based on how problems of the same kind have already been resolved whenever they have imposed themselves previously.[12] The problems thereby treated are economic problems; they revolve around how to count what is given—but not around how we might account for the manner in which we do count that which is given. The tables in which the treatment of these problems is organized must be ordered around a step-by-step procedure that will lead whoever follows it to the desired decision or solution. Such methodical, goal-oriented procedures are what Serres calls *algorithms*.[13] They spell out how to reach all the intermediary steps as one attempts to multiply quantities, to divide them, to raise them to a different (exponential) power than the one in which they appear to be given, to extract the roots of a quantity or to sum up or divide them. The overall framework of these operations, one might say, consists in finding ways of counting, as exhaustively as possible, the possibilities hosted in a quantity's reciprocal value—possibilities that are the very substance of economic thought.[14] The methods of how such a tabular organization is gained, are strictly algorithmic. An algorithm is made up of techniques or operations of how to count—what we today summarize as the operations of arithmetic. Its procedures in Babylonian science know three classes of numbers: the givens, the results, and the constants, which are the stepping stones from the given to the desired results.[15] As long as we do not attend to the possible manners of accounting for how that which is given is counted by these tables (organized into the three classes of numbers), quantities lack a proper generality; they are always concrete and singular. Generality is not seen with regard to the things given; rather, it applies to procedures only: an algorithm is an algorithm (and not an account of one's experience, like a fable or a tale, for example) because it is a general rule that can be reproduced in its experiential value by anyone who follows its steps. Once a specific procedure is put in a numerical form, one and the same algorithm can be applied arbitrarily to particular situations. Such algorithmic procedures usually end with the formulation "Behold, one will do likewise for any fraction which occurs."[16]

Against this background we can understand Serres's admiration for the anarchic reasoning that has no particular economic interest of a people at its core, but which fantasizes a reasoning proper to the world itself. The homological scenes that unfold in his homothetic space of abstraction, and that are expressed in the projective scenes of writing that accrue from it, are full of brilliance, yet the intelligence that shines in it is not that of an extraordinary priest, king, or an official expert. Methodically anarchic reasoning differs from algorithmic reasoning mainly in that it treats the *manners of accounting* where that which counts expresses its power, wittily and challengingly. The brilliance that shines in the methodical anarchic reasoning of a world that exerts itself upon itself, by double duplication, is that of a world that collects and discretizes itself in a genuinely public, and ideational, language (that of mathematics). For Serres, "intelligence is immanent and, probably, co-extensive with the Universe."[17] The world owns a huge stock in forms, he tells us. "There is a vast objective intelligence of which the artificial and the subjective constitute small subsets."[18] The new economy that corresponds to the reasoning of the world feeds from the cornucopia of ideas that the world might recognize as its own, while trying to keep track, in its reasoning, with whom and what it actually is.

Fifth Iteration (Inventing a Scale of Reproduction)

So let us turn back to Thales, and how he gradually invents a scale of reproduction for measuring the colossal manifestation of the pyramid. Thales sees in the pyramid the eminence of a principle, we said, that of homothesis. But how can we learn to en-familiarize ourselves with the meaning of this? What we can learn from Serres is that homothesis abstracts from the tabulatory accounts that preserve and collect, in their algorithmic tables, all that the gnomon indicates. One way to put it is to say that Thales steps out of the apparatus of capture's reign, and that he dares to multiply the very principle of its regime.

Let us recapitulate and see how Thales proceeds. Thales has no direct access to the object he wishes to measure, and sets out to establish the possibility of an indirect way, by a double duplication of the situation and through engendering the form of this double duplication as a reduced scalar model. He proceeds to measure the pyramid by postulating that it be possible, and elaborating his own fantastic postulation before his inner sight, that is, in theory. He begins this elaboration by building a stock of experience—a résumé—or, as we might say now, by treating what appears to be *a given* as *data to be organized* in algorithmic tables. What appears as a given, he dares to think, is given by the gnomon and can count only as indexes to something that is not exhaustively given in what the gnomon collects. This something, he considers, must be of such a *magnificent* quantity that the form of reciprocity that hosts it also hosts the size of the pyramid as one of its possible variations. If one were to en-familiarize oneself with the dimensions of the monument, and hence be capable of measuring it, this magnificent quantity is what one would need to better comprehend. Thus, after having stepped out of the immediate reign of the gnomon's apparatus, Thales gives way to a thrusting forth of his mind beyond what it is yet capable of encompassing. He wants to learn. Following Serres in his account, we can remind ourselves that before Thales will know, and be able to draw his famous circle in order to measure the pyramid, he has to iterate and go in circles, on grounds no more solid than his desire that it be possible. He has to assume a result that seems, from all he can know, beyond reach—and it is on the premise of its assumption that he must try to find an algorithm that will guide his way to the result whose solvability he presumes against all odds. This is what we have called "embracing an obstacle." In this way, Thales gradually builds up his résumé. He continuously sums up what he finds along his iterations, and attempts to treat the sums he comes up with as values proper to his hypothetical form of reciprocity of a quantity so magnificent that it hosts the invariant quantity that makes the pyramid comparable to the reduced models he is trying to build by his civically anarchic reasoning.

But what stock of experience does he draw from when attempting to build a model? Going around in his circles, Thales regards the pyramid as an objective ruler. He begins by regarding it, in the common manner of thinking, Serres suggests, as a sundial. He expects the pyramid to speak about the sun, and to indicate the hours of measuring. He marks the outlines of its shadows as time goes by, and faces a growing number of varying outlines the longer he goes on. As he continues his circles, he begins to consider all the outlined shadows (which build his stock of experience, his résumé) as variations commensurate with one another by that module of which he knows nothing more than that he must proceed according to its proportionality in his attempted act of double duplication. Thales eventually succeeds in abstracting from the idea of the gnomon, explains Serres, by changing the *real* setting of his exercise into a *formal* setting in theory: instead of bringing the pyramid to speak about the sun, he can now ask the sun to speak about the pyramid.[19] This perspective, which is now a theoretical one, no longer based on experience alone, does not, as before, require that the magnificent quantity, whose form of reciprocity hosts the invariance he seeks, be really and actually given; rather, it may remain a secret—like those secrets, inherent to materials and to tools, which forever inspire the development of a craftsman's mastership.

Hence, we can imagine how Thales's view gradually begins to change. He ceases to contemplate the variations he observes and registers, as he goes around in circles, in order to find in them a new "given," from whose concrete shape he learns a general procedure. Yet he cannot blindly compute, as was customary with the algorithmic way of thinking, what may count as constant and common throughout the transformations among all the outlined shadows. No, he begins to take the stance of the artistic craftsman—and he is well aware that what he attempts to craft must remain abstract. He sets out to craft a genuinely theoretical object, one that duplicates the objectivity of the ruler. Now, the variations begin to interest him because they must host, he thinks, the essence of an invariant quantity that, like a guest, can never appear in its familiarity as long as it is respected *as* a guest (and not subjected to the customs of one's own home). Like a guest who is familiar and strange, not due to willed disguise but by lack of alphabetized commensurability, the invariant quantity must be treated in a space, and in a language, in which the artistic craftsman too is an immigrant and a stranger. It cannot be the concrete objective space of collective memory that allows for the dramatic act of an *inceptive* conception; rather, it must be an abstract space which is capable of staging the intuitive concreteness of collective memory. From now on, Thales strives to en-familiarize himself with the immenseness of the pyramid; he no longer hopes to succeed in subjecting it to an order that he would already be familiar with. He aspires to do so, by expecting from that which changes ceaselessly (the shadows) that it be capable of speaking about what is stable in an abstract and non-concrete manner (the measured pyramid). He thinks about the setting in which he finds himself (at the foot of the pyramid, in the desert) as a formal setting, not as a real setting, and with this, Thales can find a trick to render—against all likeliness—the course of the sun *permanent*. He no longer participates in the dictates of the gnomon as a real ruler, where what it points to must belong to what is already given, but to what can be seen in what is given only by pointers to something whose magnitude is magnificent, and as such bound to remain immense, and barred from being directly experienced.

With this leap into theory, Thales no longer uses space to indicate time; he arrests time by generalizing one particular, and real, moment—that when our shadows and our bodies have the same length. As Serres puts it, he homogenizes the singularity of each day in favor of a general case—one has to freeze time in order to evoke geometry.[20] In other words, Thales must symbolize a world in which he could relate to a monument of such awesome colossality and vastness (from the Latin *colossus*, "a statue larger than life"). Like this, Thales can think with all the cunning and conquering reason of which he is capable, and yet without being disrespectful to the secret at the center of the pyramids. Such puissance is proper to the nature of intellection, Serres seems to be saying, and an intellectual nature that is not at odds with an ethics of mutual respect. We can see in the birth of mathematical theory the unlikeliness of beginning to converse abstractly.

Sixth Iteration (the Formula, a Double-articulating Application)

Thales's double-articulating application of the gnomon contemplates all possible variants of a triangle by inscribing them, theoretically, into a common compass: the course of the sun's permanency. This is how Thales eventually succeeds in conserving, in his textual formula of right-angled triangles, a universal and formal concept of similarity. Its compass is conceived by a reasoning that is proper to the world as it exerts itself upon itself—the course of the sun as collected by a duplication of the gnomon. Thales's theorem states, as a means of conservation, that if A, B, and C are points on a circle where the line AC is a diameter of the circle, then the angle ∠ ABC is a right angle.

For Serres, as we will see in a moment, recounting what Thales might have seen at the foot of the pyramid is inevitably a text about originality. Like Thales himself, Serres is not interested in revealing the signification of this origin by claiming to be familiar with it; instead, he wants to postulate, again like Thales, further theorems of universal value. Let us see what some of Serres's own postulations are, and how he sets out to elaborate on them.

The Amorous Nature of Intellectual Conception: Silent Words That Conserve the Articulations of an Impersonal Voice

First Iteration (Marking All That Is Assumed to Be Constant with a Cipher)

First we must examine the object of Serres's own double duplication. Thales, we said, double duplicated the algorithmic mode of iteration and established a textual formula that conserves an infinite amount of variations. As Thales puts the algorithmic mode of iteration in Babylonian science at stake in order to generalize from its custom, Serres puts Thales's own reasoning—which he sees as consisting in duplicating the scene—at stake in order to generalize from Thales's custom in turn. What happened in this "Thales moment" counts to Serres not so much *as the origin of geometry* (which is today's customary association with this event), but *as the inception of a staged scenery for the dealings of abstract thought*. The inception of such a staged scenery is necessary, Serres maintains, for developing proper ciphers for formal reasoning out of the formality of mathematical statements—alphabets that, like any alphabet, allow for expressing an infinity of articulations by a finite stock of elements. Thus, if Thales was capable of formulating his theorem by attending—theoretically—to the permanence of the sun's course, Serres wants to reintroduce temporality and the vividness of real happenings into the formal settings Thales established. If Thales questioned the principle of the gnomon by multiplying it, and thereby invented the space of theory (homothesis and homology, organized according to an abstract principle of similarity), Serres sets out to question the principle of theory by multiplying it, massively, as principality, and by inventing a code-relative alphabetic view on the timeless space of formal theory. Such a code-relative alphabetic view is what to him counts as the birth of physics from the spirit of mathematics.[21]

Serres's account sets out to speak about how the abstractness of an architectonics of formal ideality had been fabricated. The proposal is simple. What Thales realized, according to Serres, is threefold: (1) the possibility of reduction: Thales creates a model that extracts from the given situation a skeleton reduced from all singular context, and that is in favor of a general case; (2) Thales affirmed the idea of a module: that throughout different sizes and scales, the quantities at stake must be commensurate; (3) Thales conceived of the model in a general, not in an iconic, representational manner: he "invent[ed] the scale" of reproduction[22]—the model in general is a double staircase, a Rosetta Stone to keep the secret through inventing how to engage with it. These are the conditions that make *the creation of a model* possible, as an intellectual act of engendering. Yet, as conditions, they depend upon being bracketed and encrypted, ciphered: Thales, trying to win the immense for a mutual encounter in a realm to which both are immigrants. All familiar constancy in terms of space, time, practice, perception must be questioned and neutralized, accommodated within a cipher (containing their version's inverse, their counter direction). Driven by his desire, Thales treats them as coefficients that must, in *some* way of which he knows he can never see *how* (barely, purely, in an immediate manner), be at work within what he seeks. And indeed, once Thales comes to measure the pyramid, each condition will be raised in their powers: space will host something that does not exist, a general model; time is arrested and one of its moments is rendered perennial; practice comes to envelop not a necessity but something that appears necessary (a theory); measuring does not depend upon tactile perception but upon visual sense. Thales, in the account Serres gives of him, invented the stage of abstract conception by conquering, with neither disgrace nor sacrilege, what is, in its dignity, impenetrable: the arcanum of the pyramid's lasting and unviolated immenseness.

Second Iteration (Confluence of Multiple Geneses)

Serres's own double duplication of the Thales situation constitutes, in turn, a model. What he sees while tracing the conquering movement of Thales's act of intellection, lets him face something that appears to him as immeasurable as the pyramid must have appeared to Thales—let us call it the *graceful desire* by which he sees Thales moved. It is the desire that desires the arcanum. It is the desire for revelation of what must remain, if one does not want to violate it, concealed. So what does Serres do, in his account of Thales? He sees in the Thales situation a multiplication of originality in procedural, operative terms: *algorithmic* originality times *gnomonic* originality times *formulaic* originality times *textual* originality (the originality he adds to it when he reads Thales's story as a story of origins).[23] The multiplication of origins supports a multiplication of how we can account *with* givens by rooting them virtually, projectively, in *enciphered* constants, and by symbolically domesticating the growth of what can be yielded from these roots (the variables in all possible variation) if we carefully tend to their tabular organization. The careful tending of such graceful desire consists in treating formulaic statements as theoretical fabrics—as architectonic statues, as the silent words of a mathematical dictionary—which aspire to caress the integrity of the colossal through offering indexical manifestations of possible accounts. In such accounts, the terms of formulaic statements feature as protagonists, as actors onstage within texts of proper originality. In the plurality of such statuesquely manifest theoretical fabrics, we can render the givens comparable as things that remain, essentially, elusive and come to the world from an outer space of universal intellection. Like this, the "givens" must be regarded merely as pointers to a magnitude with which we can en-familiarize ourselves, if we collect the indexical pointers that mark that magnitude, by integrating them into a commensurate compass, stating that what can be conserved into a formula depends upon abstract conception in a realm of theory, and this realm is, essentially, inexhaustible. More concretely, in his multiplication of originality, Serres faces *an immense product*, a result that integrates the streams that spring from all these different originalities, as *the confluence of multiple geneses*.[24] The code-relative alphabetization of the ciphered theoretical space must attempt to draw balances from this immense product.[25]

So how does Serres imagine that the Greeks were able to conceive of the abstract stage of geometry? Through a fourfold genesis, he suggests: (1) a *practical* genesis which consists in "producing a reduced model, coming up with the idea of a module, tracing back what is afar to what is near";[26] (2) a *sensorial* genesis which consists in "organizing the visual representation of that which cannot be sensed immediately by touching";[27] (3) a *civic* or *epistemic* genesis which consists in "departing from astronomy and inverting the question of the sundial";[28] (4) a *conceptual* or *aesthetic* genesis which consists in "stopping time in order to metricize space, swapping the functions of variability and invariance."[29]

Third Iteration (the Residence of That which Is Genuinely Migrational)

From within this insubordinate happening of confluent streams, which Serres recounts while contemplating what Thales might have seen, Serres identifies three conditions that will firmly support to gracefully acquire a sense of inner sight (theory) by building schemata in the form of optical diagrams. Such diagrams are about a mise en scène of a form that is already here (myth, legends, several stories of origin)—they are not about the constitution of a form.[30] Nevertheless, such dramatization of form, in myth and math, contains the essence of theory, he holds, and this essence is an act: that of "importive" transportation.[31] Theory, by sending whoever reasons theoretically on travels, allows him or her to grow more familiar with what manifests itself as immense. Let us recapitulate Serres's reasoning. The sense of sight, and that which is seen, premises the following givens: position and angle, a source of light, and an object that is viewed as either dark or light.[32] The confluent streams are treated as processes of transportation, and the questions to be asked, Serres maintains, are questions of where that which is caught up in transport properly resides:

1. "*Where is the point of view?* Anywhere. Where the source of light resides. Application, relation, measurement are possible by an alignment of markers [*amers*]; one can see the sun and the peak of the pyramid aligned, or one can see the peak of the tomb and the uttermost end of the shadow aligned. This is to say that the site [of the point of view] can change place."[33]
2. "*Where is the object?* It is necessary that it too be transportable. In fact, it is transportable: either because of the shadow which it casts, or because of the model that emulates it."[34]
3. "*Where is the source of light?* It varies, it is a specific case of the gnomon. It transports the object as a shadow. It resides within the object, what we will call the miracle."[35]

It is an enchanted world, the world in confluent streams of multiple geneses, and yet it is a world of objective reasoning. It is a world in which what testifies to the immenseness of life and death can be encountered gracefully. Where a monument evokes a sense of tremendousness and seems to demand subordination, Thales shows us (through Serres) how we can en-familiarize ourselves with it by considering abstractly and carefully superordinate concepts, *hypernyms*, by dramatizing them. To conceive of the world abstractly is a form of conquering that never annexes what it conquers but "coexists" with it. To conceive abstractly *employs* (contracts into a work relationship), in an altogether original manner, *familiarity* from one's place of origin, by treating what appears to be constant in this familiarity as ciphered invariants that need to be rooted statuesquely in domains yet unknown, to be engendered by no other way than by an *anarchic* reasoning that has now, with Serres's double duplication of the Thales moment, turned *civic*. Mathematics provides keys to history, not the other way around, which means that mathematics can send us on the paths of foreign myths, legendary stories of origin which appear to contradict the ones we are familiar with. Myth transports dark contents through time, and mathematics provides the vehicles for being in charge of one's journey: this, I call "anarchic civility," the civility of a reason that acknowledges to be principled but insists that the central position can only be vicariously occupied, by substitutes; no principle, hence, is ever sovereign and beyond doubt. Such geometrical commitment to the *absence* of a leader (Greek, *archon*) by engaging with the locus of its apparent presence through seeking to provide a cipher that makes it measurable, and accountable, in general, is what I refer to here as "anarchic reason."

Fourth Iteration (Universal Genitality)

We are on a stage of abstraction, a locus in which all that features is at once native, but also immigrant to it. It is the *lieu* (locus) in which to conceive of things in their genericness, and in their universal genitality (their hermaphroditic "whiteness"[36]). It is a theorematical stage, and it enables the unfolding of plays in the scene of writing: plays that perform the measurement of originality in theory. Nothing in these plays is native to their plot lines; everything that features in them is foreign, migrant, passing through. With regard to such measurement, no one can possibly be at home when daring to make statements about what happens in a scene of originality. Such measurement depends upon one's own en-familiarization with what is awe-inspiring—on the sole condition that we can count, if only the ways of conduct are not without grace, on the hospitality of what is colossal: "The theatre of measurement shows how a secret may be decoded, how an alphabet (a writing) may be deciphered, and how a drawing may be read."[37]

In Serres's account of theory, mathematics provides keys to history, and history provides for mathematics to manifest massively: mathematics itself is historical (even though it formulates what does not extend in space and time). A scene of originality can be witnessed, Serres insists, only in mathematical language.[38] It is a scene in which something immense is posed at the discretion of a theory, and a theory is the projective articulation of an arcanum, a secret, to be engaged with. Mathematics is an anarchic reasoning that seeks to engender a circuit. Nothing more. It cannot be witnessed empirically; it can only be actualized civically. It is natural too; intellection is natural. If the essence of theory is transport, as Serres maintains, then theory is never about identifying with the revelation that takes place in abstract conceptions that are attributed to count as scenes of originality—like that of Thales and the inception of the theorem of angular measurement within a circle. It is not important whether Thales draws the circle around himself, or around a simple stick, as far as the statement of the scene in the form of a theorem is concerned. A theorem expresses a schema, an optical diagram, and the schema is a stable auxiliary construction that allows a thing to be transported. Such auxiliary constructions render all things mobile; they are vehicles.[39] They facilitate within the reality of the universal the migrational activity of that about whose essence we can say nothing more than that it is immense, a crystallization between life and death, a being about which no one knows anything beyond what can be stated of it in the universal terms of impersonal, mathematical, agreement. As a thing stated like that, in its contracted originality, one can tap into the circuit of activity that is organized in its statement. And this without, properly speaking, understanding it.

But one needs to acknowledge the theorem. And this involves, ever again, to "pay" one's coordination of familiarity, the elements of one's world, "as a tribute" to the possibility of spelling out of the theorem. That is why mathematics, to Serres, provides keys to history. What can be told by theorematical statements are projective articulations of an immense content, statuesque words that never make a sound. And in that, they are not much different from how the schemes in mythical tales work: a schema is what remains invariant, regardless of the number of times a story is told. But the schema is not the origin of this invariance; rather, it is its vehicle.[40] Every mythical tale is the dramatization of a given content. The relation between a schema, and the mobilization of an original thing that the schema affords, is essential for a tale to become tradeable (brought through tradition). Mathematics is a language, but one can speak in it only in the terms of a private, unpublished story, because what it expresses is universal *and* singular. It cannot be expressed in any other language; it can only be actualized in our native, or acquired, spoken tongues. Knowing a theorem means to have lived up with grace to the encounter with the arcanum it hosts. It can only be talked about from afar, through anecdote, on the relation between two ciphers that are, ultimately, *not* to be deciphered:

> Thales's geometry expresses, in the form of a legend, the relation between two blindnesses, that of the result of practice, and that of the subject of practice. It formulates and measures the problem yet without resolving it; it dramatizes the problem's concept, yet without explaining it; it poses the question in admirable manner yet does not answer it; it recounts the relation between two ciphers, that of the mason and that of the edifice, while decrypting none of them.[41]

Fifth Iteration (Mathematics Is the Circuit of Cunning Reason's Ruses)

A theorem renders available certain techniques, because techniques envelop a theory. They are stable coatings that package the acts of archly reasoning in scenes of originality, in abstract conception. In order to take these practices and do something with them, in order to apply these techniques, one need not *know* the theory that they envelop. But without knowing it, one does not touch upon the question of originality. It cannot be separated from the pride of a craftsman who seeks to become masterful, in the sense of conquering his material without disgracing it. As Serres puts it,

> What is the status of knowledge that is contained in a technique? A technique is always a practice that envelops a theory. The entire question—in our case that of origins—reduces here to a question of mode, the modality of this envelop. If mathematics springs one day from particular techniques, it is without doubt because of an explication of such implicit knowledge. And if the arcanum (the secret) plays a certain role in the tradition of craft, then certainly because its secret is a secret for everyone, including the master. There is a transparent knowledge that resides hidden in the craftsman's relation to stones and rubble. It resides hidden, it is locked by two turns of the bolt; it remains in the shadow. It lies in the shadow of the pyramid. This is the scene of knowing, it is here that the possible, the dreamt, conceptualised origin is staged. The secret of the builder and the stonemason, a secret for himself, for Thales and for us, this secret is the scene of shadow plays. In the shadow of the pyramid, Thales finds himself within the implicitness of knowledge, which the sun is supposed to render explicit from behind, in our

absence. Here the entire question of the relationship between the schema and history, the rapport between implicit knowledge and the workman's practice will be posed in terms of shadow and Sun, dramatization in Platonic fashion, of implicitness and explicitness, of knowledge and the technical operation: the sun of insight and understanding [*connaissance*] and of the same, the shadow of opinion, of empiricism, of objects.[42]

All things stated are objects, and objects conserve an implicit knowledge. Grasping how it is implied is the truly difficult thing, the impossible thing, because if one desires not to violate the secret, there will always be an unexplained remainder left behind. The circuit that can be established by anarchic reasoning cannot possibly exhaust its source. What reveals itself in scenes of originality, by abstract conception, is always impure. The universality of geometry reveals itself in its application, and only there. In terms of purity, geometrical universality can never be born.[43] In other words, it can never become physics, it can never be considered natural. Mathematics as a language of silent words, in contrast, allows us to consider all things natural. This is how Serres can claim that mathematics is the circuit of cunning reason, or archly staged scenes of conception. If originality is actualized in such scenes through theory, and if theory is transport and a theorem is a vehicle, then we can regard mathematical formulas as textual in a sense not unlike semiconductors are for electronics. This is indeed what Serres suggests:

Measurement, surveying, direct or immediate, are operations of application. In the sense, of course, in which a metrics, a metretics, stems from an applied science. In the sense that in most cases, the measure is the essence of application. But most of all in the sense of touching. Some unit of measure or some ruler is applied to a thing to be measured, it is laid over it, it touches it; and this as many times as necessary. Direct or indirect measurement is possible or impossible as much as this application is possible or not. Inaccessible is, hence, what I cannot touch, where I cannot lay the levelling rod, that to which I cannot apply my measuring unit. We must, so they say, go from practice to theory, by a ruse of reason, we must come up with substitute for those lengths that are inaccessible to my body, the pyramid, the sun, the ship at the horizon, the other side of the river. Mathematics would be the circuit of ruses.[44]

Sixth Iteration (the Real as a Black Spectrum)

However, to see in mathematics the quasi-electric circuit of cunning reason would be to underestimate the scope of practical activities, because the established circuit is a bridge, archly, between tactility and sight. To theorize means to organize sight according to the quasi tactility of a conceptual body that lives in the scenes that unfold on the locus of abstraction. Measuring puts two things in mutual relation, and a relation presumes a transport—of the levering rod, of the angle, of the things applied when measuring. There is an inexplicable intimacy between knowing and the problem such knowing lays out theoretically. Homothesis constitutes the locus of abstraction, and the homology—the variable equivalence—that can be expressed by the statements of homothesis belongs to the reality between product and producer. What is formulaically set up as equivalent is *an invitation to read into* what the formula states; it is not a question of addressing and answering. Reading mathematically means to stage a scene that supports trading the secret of the manifest body through scenes that are accessible only to an intellectual sense of sight. The anecdotes in which the origin of a theorem can be told imply a schema that feeds off and lives on in the dramatizations it supports. The schema, the optical diagram, can be traded only in written form. It keeps what is enveloped by practices through *not* explicating it. In proceeding like this, the schema demarcates something real, something stable and lasting that belongs to the manifest body one seeks to measure: its arcanum, its secret. And it demarcates this secret by treating it as an invariance that can only be conceived abstractly, by attributing to it a measure, as a manner of how to proceed. The stage of abstraction is the projective domain of measuring—what is being measured, by projective staging, is "the real as a black spectrum."[45] From the point of view of the craftsman who seeks to understand more about the origins implied in his material, the material's original reality resides in the shadow cast by the sun. It is the shadow that bursts with spectral information: "real knowledge of things is hosted in the essential shadow of solid bodies, in their opaque and black compactness, locked up behind the many doors and their faces."[46] Knowledge about the real is natural not despite but only because it is conceived and born abstractly. It is impure because it was conceived within the happenings of confluent streams of geneses, whose pool of possibilities is a code pool that cannot be exhausted in what it hosts. It is from the essential darkness of things that can be rendered apparent on the projective stage of abstraction, in the plays that unfold in the scene of writing, where knowledge of real things lies buried, Serres maintains. From its source springs the infinite history of analytical progress: "The body which can never be exhaustively described from analysing its bounding surfaces retains in the safe depth of the bounding surfaces' shadows a dark kernel."[47]

Remembering the stage of abstraction that supports real knowledge allows us to see the ideational purity of mathematics instead of an ideality of representations. The ideational purity of mathematics is constituted by nothing more and nothing less than the presumption that there is contained, within manifest bodies, ever more that can be explicated in theory. To see ideality in the geometrical forms, as Plato did, instead of assuming ideational purity in mathematical theorems, means to dislocate homothetics and homology into the eternity of the one moment that Thales arrested when he wished that time—the epitome of change—might speak about the solidity of the thing he faces. It means that geometry is conceived yet cannot be born. It means postulating that there is no reality to desiring conquest, that technics be either divine fate (Prometheus, Pandora, etc.) or the stigma of decadence. It holds that revelation be apocalyptic, purifying, in that it clears the spectra of recognition into the whiteness of universal genitality. This white spectrality, which supposedly allows us to recognize the identity of things as they ideally are, behind their disturbed appearance in actual existence, constitutes the idea of pure intuition. Serres's, by contrast, is an idea of objective intuition. By insisting on the essential darkness of things, Serres may sound like a worried prophet, yet it would be the prophecy of a worldly nature and a natural sexuality that is driven by the desire to conquer and master what is never intended to be a possession:

> [But] when the moment has come for this purity of geometry, inherited from the Platonic legacy, to die, when nothing can any longer be supported by intuition, when the theatre of representation will have closed its doors, secret, shadow and implication will once again explode, among those abstract forms, before the astonished eyes of mathematicians—explosions which have been prefiguring all along history, before these deaths. The line, the plane, the volume, their intervals and their regions will once more be recognized as chaotic, dense, compact … entities, full of dark pockets and secret angles. The simple and pure forms are not that simple nor that pure; they are no longer things of which we have, in our theoretical insight, exhaustive knowledge, things that are assumedly transparent without any residue. Instead they constitute infinitely entangled, objective-theoretical unknowns, tremendous virtual noemata like the stones and the objects of the world, like our masonry and our artifacts [*objets ouvrés*]. Form bears beneath its form transfinite nuclei of knowledge, with regard to which we must worry that history in its totality will not be sufficient for exhausting them, nuclei of knowledge which are profoundly inaccessible like indelible marks. Mathematical realism is weighed down and takes on that old compactness which had dissolved beneath the Platonic sun. Pure or abstract idealities cast shadows once more, are themselves full of shadows, turn black again like the pyramid. Present-day mathematics unfolds, despite its maximal abstractness and the genuine purity which is proper to it, within a lexicon which results, partially, from technology. Novel manners of listening again to the old Egyptian legend of Thales.[48]

Technology manifests as implicit ideality whose theorems are mobilized in the representations of its variables and coefficients, which are projectively staged and "importively" transported through language. Technology is bursting with implicit knowledge. Every technology is a physical "text" of statuesque words, silent but capable of sounding the world. These are texts that host accounts given about witnessing what occurs, in current states, throughout scalarly countable time—texts that host accounts of what might occur, on a projective scene of originality, one of abstract conception. And this, following Serres, is no embrace of mysticism at the expense of rigor, exactness, and reason.

5

Banking Universality: The Magnitudes of Ageing

Nothing could be more "useful" consequently than despecialized metaphysics, despite the mockery of the specialists who seem to be unaware of how destitute of future its absence would make them. Metaphysics serves to remain human and not die from it. It allows the subjective, the collective, the objective, the cognitive to survive. It ought to serve to build a new peace during these times of new war in which invisible beings put invisible beings to death. Who does it serve? At least it doesn't serve anyone, whether tyrant or guru.

Michel Serres, The Incandescent[1]

Metaphysics

These scalarities, Rosetta Stone stairways that provide bridges where there is no symmetry, and that make up the diverse homothetical stages of abstraction, capable of supporting real knowledge of the world, this is what Serres's mathematical lexicon affords to build. He refers to the organon behind this lexicality of the real world (the body-of-thinking that is of universal genitality) as "white metaphysics." The concepts this metaphysics organizes are conceived, in amorous fashion as we have seen with the legend of Thales, by the gnomonic agency that is at work in the "thought of the world." As concepts, the words of such a lexicality must be universal and therefore "white" like physical light that comprehends any color at all. These concepts, however, are not themselves forms that regulate in a definitive manner. The analysis they support is spectral.[2] They are formal in a manner that *renders* present, that *lets appear* much like technical spectra do in today's applied sciences. It is a lexicon of *operational* concepts. This chapter gives a glimpse and (an inverse) look back to the last chapters by contextualizing the developed ideas within the peculiar *organon* that supports Serres's *Lexicality of the World*. We will look more closely in order to better imagine the organization of knowledge according to a Chronopedia.

The Quickness of a Magnanimous Universe

Philosophy has to reconsider how it addresses the world, Serres maintains, and this, for him again, is an issue of metaphysics: to "categorize" literally means how to place and address (name) something properly and adequately according to "highest notions" (from *kata*, down to, and *agora*, to the public assembly). The world should be addressed as one that actively knows. The metaphysics Serres advocates revolves around a principle of invariance; operates with "white concepts" (which capture the passing of massive time, and are considered as the massively percolative, code-relative formality of spectra rather than purely ideal, immaterial forms); draws upon an anonymous, impersonal agency at work in "knowing"; and addresses the entire world only as a whole, namely, by a sixfold sheaf for its "proper name," what Serres calls "a Panonyme": one for all of its places (*Pantope*), one for all of its durations (*Panchrone*), one for the universal worker (*Panurge*, not demiurge, the public worker), one for all of the spoken tongues (*Panglosse*), one for all of knowledge (*Pangnose*) and one for all sexes, *Panthrope* (instead of only "man" as in "Anthropos"):

> The white "one" sees, if not his clones, at least his cousins everywhere. What should we call him? I name him *Pantope, Panchrone, Panurge, Pangloss, Pangnose, Panthrope*, a man or woman who's integral six times over. Far from escaping from the common run by means of exceptional qualities, she or he, hidden in the incandescence, melts into knowledge, humans and horizons. The return of the Great Pan unites them. *Nemo* becomes "one," that is to say, everybody. How?[3]

Serres thinks of himself as a materialist thinker and mathematical realist, which also means as a mystic of mathematics. We should bear this in mind when attending to the one formula that perhaps orientates his thinking as a writer and philosopher at large: to him, *reality and rationality* must be regarded as *equipollent, as equals in force, power, or validity*.[4] Relations of equipollence yield a formula; hence, they have to be written in equations, stating an identity—but they state an identity in a contractual manner. This was, at the core, Serres's argument in *The Natural Contract* (1990), which he had recently, with *The Incandescent* ([2003] 2018), extended into a detailed account of the metaphysics that can categorize such a contractual, universal, identity notion: identity, for Serres, is not subjected to form in general nor to kindred substance. It is subjected to *the material agedness* of the universe. What is at stake with this materialism is universal and generic identity, an identity that is common to all there is, but in a manner that is abundantly unsettled and restlessly active. All that Serres assumes for it, with the attribution of the equipollence relation between reality and the rationality, is that there is an *indefinite yet determinable transitoriness and transversality immanent* to such a universal identity—it is, hence, the identity of a universe that rests neither in time nor in space, but *is, massively,* the all-of-time: the principle of this universe, its law, resides in its own *active quickness*.

An equation articulated in terms of equipollence must hence be attributed a status of its own, which, for Serres, is *metaphysical*.[5] The promise of this metaphysics is not sterile and cold truth but the excitement of vulnerability, quickness, and liveliness: "White concepts form a group much more than a simple class: they proceed one from the other. Look for liberty, and you will know, look for understanding and you will invent, look for knowledge and invention together and you will not be able not to love."[6] Serres answers his own question: "What is the use of metaphysics?" by maintaining that its concepts allow us to think the incarnation of a generic "body"—a corporeality that is born from anybody's body, a corporeality for the objective sense of intuition his transcendental philosophy seeks to rediscover. Metaphysics is indispensable, he maintains, to "remain human and not to die of it."[7]

Metaphysical concepts are "white concepts" in the quantum physical sense of light's spectrality, where white is the zero-valued totality of all colors. It must be called a materialism if quantum theory is right to maintain that light is, after all, to be thought of in terms of particles, and that light comes to matter relative to how the light's radioactivity deals on its energy *niveaux*, with its charges of electricity. White light is the radiating and active emission of a sun that contracts and diffracts the *massive* passing of time. Concepts, hence, are to be thought of as statuesque spectra that facilitate percolation, and the reasoning they afford is at once "technical" and "natural." Such reasoning needs the records of measurements, accounts of what the measurements indicate and is exercised by the impersonal agency of gnomonic reason—an interplay between the neutral pronouns "it" on the side of measuring, and "one" (French: *on*) on the side of accounting, the sundial that affords to take stock of temporality in any scale, across all its durations. Whoever responds to being addressed in the terms of white concepts, whoever lets herself, himself, be addressed as the "nobody-in-particular" (French: *on*), actively so, as the "one" who is *no one* and at the same time *every one*, is the "bookkeeper" or "accountant" of this anonymous agency. This is the "announcer" (*le récitant*) of what Serres calls "le grand récit." This idea, regarding an announcement in terms of its "bigness," crystallized when preparing his honorary lecture after having been elected as a member of the Académie Française in 1990. He addressed his audience as follows:

> You are mathematicians, physicists, astronomers, biologists etc., you have taught me all that I know, but ultimately, you taught me but one thing. This thing, entirely new and precious, you all have invented it at almost the same time! And this is what unites you today and what permits me to address you all as one single person, as if as a single collective. Today, for the first time in history, all sciences have learned to date their object. The astronomer knows how to date precisely the big bang, the galaxies; the white dwarf or the red giant has a precise date of birth; the biologist knows how to date the birth of species as well as that of microbes; the geophysicist knows how to date the age of the Earth; the paleoanthropologist knows how to date each of the hominids that appeared before Homo sapiens; the linguist knows how to date the birth of each of the languages he studies, etc. … I see you, and yet I see but one and the same person. You. And with one stroke, the total spectrum of history is laid out before our eyes.[8]

The world. Generic identity, restless unsettlement, anonymous agency. The auto-logos of a world that reads within itself in an active manner that is called "connaître," "knowing"—does this not announce *the very end of metaphysics*? How can such a philosophy possibly remain committed to a *dialectics* if we read this term literally, as a reasoning that presupposes a kind of lucidity that shines through, "dia," while being realistic enough to take into account the fact that perfect transparency is an idealization? Let us grant dialectics for a moment the status of the very motive force, the *tragic plot* whose drama unfolds and inhabits the grand ambition of classical metaphysics, and that prevents the latter from ever realizing and fulfilling an ideal of its central promise, namely, to deliver a thinking that is of *autochthonous* roots. A thinking that "roots" on high, in the sky. How can Serres's thinking be committed to a reasoning that does not forget that it always only produces imperfect insights, that these insights can never rid themselves from being, to a certain degree, confused, occluded, cloudy rather than bright and clear, that the ideal at work in metaphysics' postulated universal commonality is bound to always already be corrupt, impure and invested with personal interests in domination? In short, is not the postulation of this identity in terms of equipollence between reality and rationality the ultimate collapse of reason itself—both dialectical reasoning and metaphysical reason?

Invariance: Genericness in Terms of Entropy and Negentropy

We need to consider what it entails for Serres to say "information theory is the philosophy of physics."[9] As we saw in Chapter 2, "Quantum Literacy," Serres maintains that it has been a mistake of information theory prior to Brillouin to claim as its own the principle that governs thermodynamics. This principle is that of "teleonomy"—a *nomos*, a lawfulness, that acts from a distant point at the end or at the beginning of time. Serres complements teleonomy with another principle: "invariance." Faithful to his model of percolative time, Serres's metaphysics maintains that this *nomos*, which acts from a distance, is determining all that happens while being, itself, undecided and receptive to what Serres calls "temporal transcendentals."[10] Thus, *invariance*, to him, takes precedence over models of this *nomos*. To him, it is the legitimate principle of information theory. Let us try to understand why.

It is crucial to grasp what is at stake here. Without going very far into technical details, let us remember that entropy in thermodynamics entails an ideal order attributed to a "system" of which it is assumed that it be infinite. But "an infinite" system is not countable, and hence cannot be regarded as a system—not without making some further assumption as to a limiting function that cuts through this infinite. In thermodynamics, the operationalization of this idea assumes that the amount total of energy in the universe is: (1) finite; and (2) a terminable invariant. Its magnitude can neither increase nor decrease. Entropy, at the level of the laws of thermodynamics, is not an operable but a descriptive term for the state in which every one of such a *totality of possible events is equally likely to happen next*. This state is the ideal called "thermodynamic equilibrium," and heat—or rather its measurement in temperature— is the agential operator in it. Any real system in this state would have disintegrated and dissolved all forms of organization, which is what worried some people in the nineteenth and early twentieth centuries. They speculatively considered that the (then) novel laws of nature would *by necessity* end up in the "heat death" of the universe (Baron Kelvin, Heisenberg, Rankine). In a universe of maximal entropy, there would be no life.

Now to the crux of the story. The father of thermodynamics, Nicolas Léonard Sadi Carnot, formulated the first law as an abstract cycle, and the problem that features in this speculative end time scenario was how to "reason" the abstractness of his model when rooting it empirically. Carnot's cycle models an ideational condition—a motor, as Serres thinks of it[11]—while of course in practice, the laws of thermodynamics are applied to systems that derive from this ideational condition, to the metastable balances they maintain among each other. And here, in order to allow for their empirical description, the principle of invariance is usually translated into rules that render constancy. But there is a crucial mathematical distinction between invariance and constancy: invariance, in contrast to constancy, does not require any a priori specification of that peculiar quantity. If we speak of invariance, we argue with *algebraic elements*, in terms of *equations and their immanent transversality*; if we speak of constancy, we argue on the level of *functional mappings* that trace some of this "transversality" in empirical observation, which they then render explicit. It is the functional mappings that give us control and praxis, which frame the conditions of possibility. These framings, however, are empirically motivated, rooted in a representational paradigm: they must presume the identity of the elements, and hence disconnect the framed conditions of possibility from the source from which they are extracted when being framed. Serres's critique is that as long as heat is the operational concept, not much is lost when the difference between invariance and constancy is neglected, because heat counts as a continuous force; all particles in a physics of forces are rightly thought of as equivalents. But when information rather than heat is the operational concept, as is the case in information science, this distinction becomes relevant: unlike heat, *electricity is not a continuous force but electromagnetic, quantum physical*. Its quanta must be regarded as being at once continuous and discontinuous. There is an decisive role played by code in electricity. In quantum physics, hence, constancy preempts the elements of a modelled system of their virtually equivalent shares in an immanent transversality. This immanent transversality is never exhaustively grasped in the way a particular functional rendering of that system arrests its undecided matrix of conditionability (a generic identity not subjected to any particular order relations).

The implications are weighty. With *invariance,* we always think in terms of the greatest possible preservation (of immanent transversality), while with *constancy*, we always think in terms of the most reasonable expenditure. Now, as long as physics thinks of itself as the "other" to life, and as long as the focus is mainly *analytical*, this distinction might have seemed unnecessarily moralistic. But as soon as the focus is *synthesis*, in chemistry and also in the study of biological systems—organisms—it turns substantial. We have seen how Schrödinger introduced the notion of *negative entropy* in order to distinguish animate systems from inanimate ones by saying that animate systems are capable of binding and incorporating a kind of energy, which he called "free" in the sense of "available," or "unbound." *Organisms import negentropy,* as he put it, *and the more they do so the more they rid themselves of entropy, that is, of undecidedness*—in other words, life forms minimize entropy and maximize negentropy.[12] Biology and physics are thereby placed in a competitive relation about who has the last say with regard to the ultimate reservoir of resources. Specific natures (biology) and universal nature (physics) are arrested dialectically in a logical opposition: from the point of view of a principle of teleonomy (which operates with constancy), specificity appears to consume universality, and universality appears to consume specificity. The pragmatic answer to this dilemma has been to revert to notions of "norm" and "normalization." But from a perspective of principles (with Serres: metaphysical), identity (local, specific) henceforth seems to negate genericness (universal), and genericness (universal) seems to dissolve identity (local, specific) into an undecidedness (global, general) that is, ultimately, hostile to the rich variety of life forms.

So how does the notion of invariance play in here? It brackets the question of the "free energy's reservoirs" finitude, and treats the quantity of this finitude algebraically as invariance, meaning at once indeterminate (not subjected to ordering relations), but determinable.[13] With a mathematical understanding of invariance, we can link identity and genericness in terms of the scalarities of the universal instead of opposing the two as incommensurable in the scales of empirical observation. *Thereby we shift focus from a preoccupation with expenditure to one with preservation. We keep the level of morality apart from that of science's claims to universality.* As Serres puts it, it liberates us from the reign of an anonymous regime of thanatocracy that came to power on behalf of a betrayal: by declaring that it protects life-in-general, it reigns by actually administrating our relations to death[14]—which, to Serres, are the source of forms. In relating life to death, the rational can learn from real singularities, rather than dominating the real.

Genuine to and Immanent to the All of Time: Le "logiciél intra-matériel"

Universal *massive* invariance, in Serres's metaphysics, is neither biological nor physical nor chemical; rather, it is *philosophical*. This is how he can say that "information science is the philosophy of physics."[15] Let us look at this idea more closely.

Brillouin, who like Monod is one of the key influences in Serres's understanding of communication, generalized Schrödinger's notion of negative entropy from thermodynamics more strictly, and applied it to information science (as discussed in Chapter 2). To recapitulate: instead of restricting use of this distinction (entropy/negentropy) with regard to energy, where it amounts to deciding about *free and available energy* versus *bound and distributed energy*, Brillouin applied it also to information, where the boundedness versus unboundedness relates to *the amount of information conserved in sign chains* that circulate in channels of communication. A message with little ambiguity has high negentropy, whereas one with much ambiguity has high entropy. It is important to remember here that in information science, as opposed to thermodynamics, entropy is *not a descriptive term* but *an operational one*: entropy acts as a measure of order only because this order is set relative to a channel.

From a philosophical point of view, however, does this generalization not suggest that Brillouin thereby voided the original commitment of entropy theory to realism, delivering it to a framework of linguistic transcendentality? With Serres, we would be mistaken to think so. Communication, for him, means to trace back the thermodynamic force (heat) to its entanglement with code: "communication," for him, means *the material exchange (by import and export) of quanta of electrical charges*—quanta that must count, as we know, as at once discrete and continuous, particle and wave, magnitude and code. Such exchange happens on small scales (his notion of *softness* as a category, as in software ("logiciél")) as well as on large scales (his notion of *hardness* as a category, as in hardware ("matériel")). There is, hence, a materialist point of view from which an apparently animistic statement like "the world exerts itself upon itself" and "the world renders itself visible to itself, and regards this rendering of itself: Here resides the meaning of the word *theoria*" (cf. Chapter 4) does not amount to a dualist metaphysics where a distinction between subject and object is always already presupposed. It does, however, introduce a *transcendentality*. But the a priori of this transcendentality is not one that applies to an individual subject's absolute faculties (unlike Kant's a priori of the forms of intuition). It applies to the generic subject of the white concepts—*Serres's transcendental is objective*, structural and multiplicitous. This generic subject, as we saw, is organized in terms of groups (in the mathematical sense of "group"), introducing local variety without a given (a priori) integral.[16] The transcendental is thereby handed over from the domain of a subject (as we know it from modern philosophy) to the domain of objects.[17] Objects, too, acquire a novel status: there is a generalized variation, a worldly (*la variation généralisé, la variation mondiale*) variation that exists "as an objective support for an information to be received, conserved, emitted."[18] The impersonal cogito of the objects, a cogito that is distributed throughout all the things in the world, is *also* a *nomos* that acts from a distance. Serres thinks of it as "un type de logiciél intramatériel," as we have discussed in the last chapter: "This type of intra-material software conditions our cognitive performance, as if it were a kind of transcendental objective."[19] Even though it is an impersonal cogito, the neutrality of its impersonality cannot be taken for granted but is one that needs to actively be considered: this is what Serres's sixfold proper name for the world, constitutive for the structural groupings (rather than classifications!) his white metaphysics is to grant.

Serres's *nomos* that acts from a distance, his "transcendental objective" (goal, telos), now depends upon being itself *instructed*. The telos of this *nomos* does not only act determinatively from the end (or beginning) of time and upon all that can (have) happen(ed) until (since) then, but it itself is being acted upon instructively, from all that happens within the *agedness of the universe*: it *prescribes* as much as it *is being instructed*. The objective intelligence Serres affirms is not the intelligence of an individual, personal subject. It is distributed among all the things in the world, and it reasons in a twofold manner: by algorithm (mechanism) and mnemotechnics (graphisms). Serres's objective transcendentalism responds to how the universe, in terms of its scalar categories (softness and hardness) can learn, and forget. Let us recall: The gnomon *takes stock of measured temporality*, but not in the sense of keeping track of history. It does not take stock of temporality in a sense that would seek to *describe its passing immaterially*. But nor does it do so in a manner that would seek to program it with will and intention. Serres stresses the importance of Brillouin's generalization of negentropy, because, for him, when the gnomon or any other clock measures temporality, it imports *unbound quanta of temporality* from what is the entirety of time (the universe) just like a plant imports and organizes *unbound energy* from the solar light. But unlike the plant, the clock does not *metabolize and organize* these quanta of energy: Serres maintains that it *banks* its unbound quanta of temporality. Such a taking stock of temporality is the kind of writing with a stylus that needs no hand to guide it. Its concepts are statuesque spectra that facilitate the massive percolation of time, and the language at work in such concepts is that of mathematics, which is, and facilitates, the circuit of impersonal cunnings. The world speaks, and all things articulate in the silent but objective language of mathematics when they exchange, store, process, and receive information among each other.[20] The concepts of Serres's new philosophy are the amplifiers of this massive rumbling. They act a bit like radio channels and loudspeakers—in fact, like all artefacts, which Serres aptly calls in French "des objets ouvrés" (crafted, but also opened-up, objects).

Hence, mathematics provides, first, entries into a novel kind of lexicon, rather than being a support or a guard rail.[21] Its concepts are of "invariant formality," silent spectral sculptures that manifest "the any form in general" [*la morphé en général, l'ensemble idéal de morphés singuliers*], as Serres specifies.[22] The impersonal cogito of his "intra-material software" produces a logos that speaks silently, physically and neutrally (the gnomon is of neutral gender—yet neutral in a sense that must be actively and *situationally* achieved, through providing, algebraically, contractually, always the *inverse* to any positive or negative).

White Metaphysics: How Old Does the World Think It Is?

With this *banking of unbound quanta of temporality*, Serres's metaphysics operates with a notion of "neutrality." We have to understand the full implication of Serres's replacement of teleonomy with invariance, and of his "objective sense of intuition" that supports a "neutral logos"; the important thing is not to confuse this notion of invariance with identity, as we saw. Identity needs to count as something substantial. How to think about such "substance"? It is not merely a slippery metaphor when Serres reverts here to finance. Thinking of equations in terms of equipollence (equality in terms of Serres's triple tress of force (energy), power (form) *or* validity (information)) amounts to a metaphysics of restless, massively active—*ageing*—universal identity. Serres can be read in relation to Friedrich Nietzsche. His universe as the all-of-time challenges the latter's doctrine of the eternal recurrence of the same. Serres's metaphysics of complex, polylateral valuation allows for extrapolating *a nature of economy*—in a similar gesture, and doubtlessly tackling the same problem (that of evil), as Nietzsche's metaphysics of revenge allows for extrapolating *a nature of morality*. Who does the earth think it is, Nietzsche asked in order to characterize and dramatically address the dawn of a new world age that he saw breaking. Serres poses the question: How old does the world think it is? His *nature of economy*, derived from a *metaphysics of valuation* (categorization in terms of white concepts), introduces a medium for "currencies" into Nietzsche's cycle of returns. Instead of the ultimate temporality of an eternity, Serres is ready to consider the *tempers* of time: the neutral element, which makes the concepts in the mathematical lexicon "white," turns the scales of natural magnitudes into currencies that can acquire more or less capacity. Nature, for Serres, is *magnanimous*, from *magnus* (great) and *animus* (mind, soul, spirit). It cannot merely be measured in terms of *magnitudes*, employing the etymon *-tudo*, a suffix forming abstract nouns from adjectives and participles. Serres's nature is information-theoretic, quantum physical, *quick* and *spirited* (no longer thermodynamic, as Nietzsche's arguably was).

The medium for currencies Serres introduces is provided by the *neutral elements* his "white metaphysics" requires in order to address the *agedness* of the universe.

These elements are not neutral because they are normal, stripped from all properties other than a general base. Rather, they are "neutral" because the axiomatics—the systems of valuation—that build on them have to account for all of the properties that might be attributed to the world. Here we understand clearly the implications of Serres's relation of equipollence: if the real and the rational are to be regarded as equals in terms of power, force, or validity, calculations with such equations are not to settle with leaving any rest; that is why the idea of a natural contract considers nature as active and restless.[23] The neutral elements are elements of an "omnipotent (and hence impotent) neutrality." They cannot be taken for granted (they are what needs to be "achieved" via a realist materialism of identity/equations as stating relations of equipollence), whereas value, as the invariant magnitude (unordered, material), the amount total of such omnipotent neutrality, must be taken as given. This is what "metaphysics of value" means.

If the entropic universe is a universe in which all things are at once nothing-at-all and anything-at-all, then the cogito at work in universal reason can no longer feel entitled to address the world in any immediate way. There is a chapter in *The Incandescent* entitled "Access to the Universal," which begins by maintaining that, hence, we have to reconsider how we address the world. In it, Serres complements the *logical ladder of differentiation and speciation* (which proceeds by identifying a common denominator) with what he calls an *entropic ladder of dedifferentiation and neutralization* that proceeds by identifying a common factor. By replacing teleonomy with invariance, Serres's metaphysical universality is not one where one looks for necessities; rather, it is where one turns and finds only possibilities—out of which accrue regularities, temporal orders that save (bank) time. Hence, there is authorship in Serres's metaphysics, despite the anonymous and objective agency at work in it. *To author*, for Serres, means *to augment*.[24] Access toward the universal grants augmentation. But how can augmentation mean something else than generalization, advances on an orderly and logical ladder whose steps are deductions? If Serres's philosophy is a realism, and not an idealism, as he claims, then there must be another way to think about "augmentation." But if things in their universal genericness are nothing-at-all in a manner in which they can be anything-at-all, then surely this "augmentation" cannot be concerned with the universal nature of things either, or can it?

One way out would be to assume that this metaphysical status of value, and the nature of economy at stake, is a transcendent—fatalist—*government of pure capital*, of competition within a materialism of general equivalence *that has lost all reason*. But Serres's philosophy is a natural philosophy, not a political philosophy (although, of course, it is political too; we must relate this specification with regard to the academic background Serres is coming from, cf. the introduction). Serres's materialism is one that *quantizes the agedness of the all-of-time* (I will turn shortly to this in more detail). "Capital," if we so call the magnanimous magnitude of the invariant amount total of massive value in Serres's metaphysical universe, is not the positivity of an empty form—it is *the plentiful abundance of a void that is incandescent*.

But first, how can we think about augmentation? We must turn again to this notion of *equipollence between rationality and reality* at this point, and ask about that peculiar status of such equatorial, scalar equationality that facilitates counting and measuring time in a manner that regards its percolative, massive, passing. It is metaphysical, we saw, but in which sense? If the *physical* nature of things is universal, then what exactly is the concern of metaphysics? It cannot only describe the universe, which would render it absurd to hold on to a notion of authorship—unless one were to postulate history as this author. Serres clearly declines to do so: to conceive of the entropic universe in terms of invariance instead of teleonomy is to *dethrone* history from exactly this position.[25]

The status of an equation is metaphysical insofar as Serres's metaphysics affirms one operative law to reign universally, *the law of chance* (amounting to the assumption that the principle nature of the universe [rational] is entropic, while the aged nature of the universe [real] is negentropic). The status of an equation is therefore universal, insofar as it *contracts* what this law states: all proceeding steps (as far as this metaphysics' universality is concerned) are to be equally likely to happen in any moment. Serres's metaphysics, then, is *the metaphysics of these contracts*, and the materialism of identity it entails makes it a metaphysics of the impartial nature of law. Serres is thereby proposing a metaphysical notion of freedom. *To augment means to proceed on the entropic ladder of neutralization,* because steps that identify a common factor (by seeking to identify a neutral element rather than settling on a common denominator) render in multiplicitous manner how a bond can be decoupled, and hence guarantee the articulation of contracts *to continue at any point in all directions.*

Freedom

If then there exists a class of white concepts, whose indefiniteness distinguishes them from the concepts defined and refined by the sciences so as to make them falsifiable and operational, Freedom belongs to it. How many women and men live free? Slaves of a party, of an ideology, if it's a question of politics, of societal conventions, of cosmetic or intellectual fashions, of any pressure group in which clones surround a perverse leader, of voracious appetites disgusting to others, of an organized network in which the paths always lead somewhere, would they agree to pay the price of a free life with open relations?

Michel Serres, *The Incandescent*[26]

The Neutral Element: Materialism of Identity

Generic identity is metaphysical and universal for Serres in the sense that *to be* is not just *to be the value of a bound variable,* to quote the famous formulation by another so-called computational metaphysician, Willard Van Orman Quine. Quine's ontological notion of *being* all too readily sacrifices the possibility of Serres's entropic ladder that proceeds not only toward differentiation but also towards dedifferentiation and neutralization. Hence it also sacrifices the dimension of a metaphysical notion of freedom. For Serres, then, *to be*, in his generic sense, means not only to figure variably but also, more profoundly, to be abundantly unsettled and restlessly active. Serres proposes a materialism of identity in which incandescence is the universal property, the common factor that affords an objective sense (think: direction) of intuition. His materialism of identity conceives of its *incandescence* as a *fragmented and distributed elementaricity of discretely packaged indefiniteness*—like the energetic traces of cosmic dust in the impure vacuum of the outer space, which astrochemists today talk about, and from where Serres, without a doubt, imports the concept. Incandescent materiality can not only enter into bonds, by exchanging energy packages, but also is capable of nuclear synthesis and nuclear fission. Like today's astrophysical universe, Serres's metaphysical universe is not only dynamic and cyclical but also is expanding.

Thus, much of what is needed, for making sense of Serres's proposal, can be gathered in the question of how we think of the status of a formula. Does it really *state* identity? Does it *express* it? Does it *confine* it? Does it *realize* it? Does it *signify* it? Does it *refer* to it? For Serres, it does all of these things, but only mediately so: a formula is a *vehicle to travel through the immanent versatility of universal activity*. Whenever a formula is transcribed into a *functional mapping* that singles out a particular angle of the quarrelsome noise[27] at work in the immanent transversality of such generic identity, it creates a channel of transit. It *takes* place, it *conceives* space and it *excites* the neutrality of the incandescent void. It renders the universal graspable, in singular ways (we will discuss this relation between universality and singularity in Chapter 7). Every functional mapping plots a solution of the enigma that the universe *both is and is not* "one"; likewise, the transcription of a formula into an *algorithm* explicates the mechanisms with which such a re-solution can be generalized and differentiated.

A formula, for Serres, *dis-ciphers identity as a way of re-solving it.* We would be on the wrong path if we simply concluded from this that Serres affirms obscurity at the heart of knowledge, that a formula for him conserves an insuperable sacrality, that its formality guards devotedly a secret that is not ever meant to be shared in all the time of the world; that formula would merely deprive whoever attends to them from a nucleus of bare truth, through distraction, concealment, camouflage, or dressing up. Rather, for Serres, a formula *dis-ciphers identity* by providing numerous keys to encrypt and decipher it—it provides these keys in the code of just such "transcriptions" that render equations re-solvable (functional mappings) and its solutions computable (algorithms). It is the dis-ciphering of identity that gives us at once theory and praxis, without giving us absolute legitimization in either one domain.

But what exactly should be attractive about this idea of *dis-ciphering*? It is from such transcriptions that render identity in re-solutions that Serres's universality can be one that expands—despite it being, from the very beginning, itself and only itself, *as all that it generically is*. Serres's universality is actively comprehending itself, and as it reaches longingly ahead of what it can already grant, and strives to comprehend more of all that it has been encompassing all along, it expands in magnanimity.[28] This arguably mystical idea backs up Serres's central affirmation, which he shares with Nietzsche: *meaning originates in promises.* For Serres, I want to suggest, the metaphysical status of equations in terms of equipollence is exactly this: a promise. But it is a promise, however, that cannot be made *without having already been, in ideation, where it will lead whoever follows it*. A promise that is a well of sense and knowledge like this decouples credit from capital and reverses the customary vector: in order to make it, one does not need to own capital and give credit; rather, one must, quite inversely and daringly so, take credit from a source whose solvency is all but granted. Meaning so conceived is neither public nor private, initially. The promises from which it originates do not take some of an originally public meaningfulness away, lock it up and make it proprietary. It is only through encryption of identity, and the rendering of the code of that encryption into a script (its alpha-numerical-ness) that both are coming into being. A code must be made explicit in order to be shared (hence it establishes publicity), but those who know the code build a sphere of *privacy* against those who do not.

Dis-ciphering and transcription are how Serres's entropic ladder of augmentation works. They are what Brillouin's principle of negentropy, when applied to information (rather than to energy) afford: here, Brillouin's negentropy does not fix order into an organization, like Schrödinger's negative entropy does. Brillouin's provides the keys with which an order locked into an organization can be unlocked: dis-ciphering and transcription afford to complement the evolutionary ladder of differentiation and specialization with an "involutionary" one of de-differentiation and neutralization.[29]

The truly metaphysical question now is whether there is an ultimate limit to the entropic ladder: "*Does there exist a marker, a limit below this dedifferentiation into neutrality?*"[30] For Serres, the prefix "meta-" means exactly this: *ce seuil là*, this bottom limit. It does not signify, as has often been maintained, *above* or *beyond*; rather, it means *below*. For him, yes, this limit does exist: *metaphysics itself actively de-signates it*. Metaphysics is the mathematical group of all white concepts.[31] It is this "group" in the mathematical sense of the word "group": a group comprehends a computable solution space; the concept of "group" has introduced the structural point of view to mathematics (Évariste Galois, Niels Henrik Abel, Richard Dedekind, among others). Hence, Serres's metaphysics is committed to structure, but not in the sense that it would look for one universal and fundamental structure that accommodates things in an orderly deductive, or inductive manner. To him, the order of a structure is an "island of rationality" in the oceans of noise.[32] We will come back to the relevance of structure for Serres's approach to history later (in Chapter 7).

With this idea Serres is actually very much in tune with the rising interest in the philosophy of mathematics. Fernando Zalameo has, in a recent book entitled *Synthetic Philosophy, Contemporary Mathematics* (2012),[33] urgently called for a philosophy that engages with the levels of abstraction that "real" mathematicians have worked with since the introduction of the group concept, and especially since the 1950s: category theory, sheaf and topos theory, among many other branches of various algebraic geometries, topologies, co-homologies, all treat objects as if free-floating, without anchor or gravity, without a total integral, but with the capacity to be "glued" or "bridged," as mathematicians say, from wherever to wherever appears as full of promise.[34]

(Pan's) Glossematics: The Economy That Deals with "Purport"

In other words, what we have to consider is the decoupling of code from signification. We have to consider *the a-signifying character of mathematical notation*, in the sense of a notation that is decoupled from the substance/content it forms/expresses. This is exactly why we need to never confuse "invariance" (indexed by code) with "identity" (substantial, abundantly chaotically (noisily but beautifully) significant). We can refer at this point to Louis Hjelmslev's concept of double articulation,[35] but not in order to subordinate such a mathematical notion to linguistics. In fact, Hjelmslev himself is perhaps the most eminent scholar who teaches us otherwise, namely, the exact inverse; the notion of the double articulation allows him to *treat linguistics as subordinate to algebra*—but there is good reason why he would not be well considered as an analytical epistemologist or a logicist. His relation between algebra and linguistics is fashioned in a way that sidesteps the analytical paradigm's supposedly necessary mediation by a logical calculus (the analytical paradigm in twentieth-century philosophy). His approach to language looks at it instead in terms of *a natural economy*. It studies how languages produce significance from circulating a linguistic kind of indefinite unit, a linguistic universal which he calls "purport." Purport is the invariant quantity that transits through the domain of language and characterizes its economy, just as in thermodynamics, energy transits through the universe and characterizes the universal nature at stake in physics. The crucial assumption for both is that the total amount be taken as indefinite but invariant (it can neither decrease nor increase). This entails that the quantities secured under such invariance be counted neither in terms of *enumeration* nor *measurement* nor immediate *accounting*, but in the *interplay* of all three as set up and governed by an alphanumerical calculus (the code of a cipher, an articulate zeroness, as in Chapter 2 in this book). If the a priori in Serres's transcendental is universal in a structural sense, it is because arithmetic itself is still regarded as universal but also as structural in the sense of such alphanumerical encryption. Arithmetic is algebraically conditioned; this is the core idea behind the turn to universal algebra in the early twentieth century.[36] For contemporary mathematics, *an* algebra is precisely this: *an* arithmetic, both with indefinite articles. Every philosophy of computation that follows the Church-Turing paradigm of computability ignores this very important fact.[37]

To maintain this indefiniteness is the precise aim of algebraic treatment of "an unknown" *as* an unknown in both axial aspects, the *quantitative* ones as well as the *qualitative* ones. Hjelmslev's purport in his glossematics treats meaning in the same way the physics of heat in thermodynamics treats energy, or as number theory treats the real numbers: as ubiquitous, indefinite and generic, "an unanalyzed, amorphous continuum."[38] This continuum is itself ungraspable except via a particular organization that is imposed upon it in a particular language (or subsystem or calculus, respectively for thermodynamics or number theory).[39]

Serres's notion of incandescent neutrality dis-solves the thermodynamic model to the level of quantum physics, where accordingly, this "unanalyzed, amorphous continuum" is subject to a kind of quantization, the "quanta" of which are undecided as to whether they must count as discrete or continuous, wave or particle, magnitude or code. We are obliged to always take both aspects into account. The most important implication of this is that the notion of entropy changes. While the Hjelmslevian purport manifests immediately as the spectrum itself, superimposed on which the different languages organize zones according to their own various and variable manners (rules, grammaticality), Serres's background of incandescent neutrality decouples Hjelmslev's "purport" from any the immediate givenness of a spectrum. Energetic entropy is complemented with informational negentropy, and spectra reveal themselves as "conceptions": Glossematics, for Serres, is pan-glossematics, and it is this aspect which makes his philosophy a *realist philosophy* that is capable of providing for a science of history as well as for a history of science.

Serres's universal intelligence, and the objective agency at work in reason when white, spectral concepts are in play, produces a logos (all that can be articulated with the lexicon of white concepts) that cannot dispense with a metaphysics, nor with categorical registers determining how that which is universal in all things can be "identified."[40] These categorical registers, to Serres, cannot be *deduced* from the principle law that reigns in his universality, the law of objective chance. Rather, they must be *extracted* from the name of the world, the name that impersonates the total amount of all that belongs to the world.[41] The world, then, is the only referent to a "proper name" and "address" in his metaphysics. The world, therefore, must be attributed a legal status (the *natural contract*, the insatiable ("panurgist") articulation of universal rights for all things of the world[42]). This is the motivation to assign to the world by the sixfold proper name (from which all categorical registers in Serres's metaphysics are to be extracted): *Pantopy* (all places and modes of placing), *Panchrony* (all durations), *Panurge* (the universal worker), *Pangloss* (all spoken tongues), *Pangnosis* (all knowledges), *Panthropic* (all sexes).[43] But even if the world can be addressed as a legal subject, its "panonyme" is still affected by all that pertains to it: *Panic*—which in its Greek sense means "all that pertains to Pan," the god of woods and fields, the source of mysterious sounds that cause contagious, groundless fear in herds, crowds or in people in lonely spots.

Quanta of Contemporaneity: Heat to Incandescence, Storage to Bank Account

Serres references the notion of the incandescent to Georges-Louis Leclerc, Comte de Buffon, a French naturalist, mathematician, cosmologist and contributing author to Jean le Rond d'Alembert and Denis Diderot's encyclopedia in the eighteenth century. Buffon wrote a *Natural History of the Earth* in several volumes (1797 to 1807), in which he criticized Carl Linnaeus's taxonomical approach to natural history, outlined a history of the earth with little relation to the biblical account, and proposed a theory of reproduction that ran counter to the prevailing theory of preformation. Serres writes,

> Buffon heated clay balls mixed with iron to incandescence and then let them cool down in order to calculate the age of the planet from these reduced models. Neither Newton nor his universe of forces has any memory; Buffon's burning hot balls accumulate energy in the form of heat and therefore function as bank accounts spending their money as they cool down; here is a new clock. It doesn't count the Newton-style reversible time indicated by my watch but the irreversible and entropic time of the wearing out of its parts, therefore of its ageing.[44]

It is this notion of *an ageing of the earth* that is central to Serres's philosophy. It is a notion which has undergone a marked evolution since Buffon, and which perhaps culminated in Nietzsche's conception of a nature of morality, and in Zarathustra, the advocate-prophet who announces the coming of a new people for a new age, an age of the eternal return of the same. Serres is well aware of this background. His book *The Incandescent* can count above all as a book on humanism in the terms of its "coming of age." He picks up the Enlightenment gesture of factoring-in a subject's age into how it is to be addressed before the law (in German: *Mündigkeit*), but not in order to directly continue with that tradition of humanism. Rather, he seeks to "geometrize" such factoring-in of age; the kind of "ageing" at stake when conceived through material incandescence not only pertains to individuals as political subjects. It provides different kind of clocks, he insists and asks: What would be the gnomons in such clocks? His entire interest in a humanism as "hominescence" is seeking to answer this question. Hence what Serres takes and augments from this idea of a natural philosophy is a *material* notion of agedness (*une vieillesse*) in which the human partakes among all other things in this world: The way in which time passes massively is thereby brought in a *mutually implicative relation* with the way time comes to "act," together with space and mass, as history; to the *quickness* of an incandescent universe corresponds to the intelligence that, for Serres, must be regarded as coextensive with the universe.

Serres is preoccupied, as Nietzsche had been, with how *to deliver the human from the spirit of revenge*; hence the importance of Pan, as the mythical persona that links the grand scope of a metaphysics with the possibility of a state of panic. But Serres rejects Nietzsche's doctrine of the eternal recurrence of the same. Nietzsche's model arguably roots in the then contemporary paradigm of physics as dynamics (and hence, as governed by a calculus). Nietzsche's model seeks to contradict its total reign, by complementing a notion of nature in physics with one in morality; but arguably it still presents just a negation, not an inversion. Serres's model roots in a quantum physical paradigm, where mechanics is again prior to dynamics. Everything that happens here happens within the active ("quick") communicative transversality immanent to a substantial all-of-time. While Nietzsche raises "revenge" to the status of a metaphysical (substantial) concept, arguably in order to dissolve the cyclical directionality it implies into a material state of in-determined-ness, Serres does the same with "value": value is a substantial (metaphysical) concept for him, and it is dissolved into a material state of in-determined-ness because it corresponds to a materialism of the "Real Age" proper to all things universal. In Serres's equipollence notion of identity, *age is incandescently indeterminate and active.*

Nietzsche's unsettling dictum was that all meaning originates in promise and derives from a relation of debt, from which it "naturally" evolves according to the favors and privileges negotiated in the course of settlement. The agency at work in such a promise is shifted over by Serres from subjectivity to objectivity. The transformation of Nietzsche's dictum hence reads as follows in Serres: "All things, in principle, behave like memories. The universe banks accounts. All things are of number, the memory of the world conserves traces."[45] "Agedness" is what the universe banks on and accounts. The traffic that circulates in the immanent transversality of the universe—within an all-of-time—counts for Serres as *transactions of incandescent quanta of this agedness.* In other words, "value," raised to its metaphysical level through identity in terms of equipollence, is "capital" *before it can be attributed* (from Lat. *ad-* and *tribuere,* for assign, give, bestow to) *to things* as their particular "proper" value. It is capital insofar as the universe actively banks quanta of its own agedness, in deposits that are accumulated and spent in how "things behave like memories." The "transits" that circulate in this immanent transversality—within an all-of-time—are not *trans-itive*; rather, *they are trans-actional.*

Thus, Serres's universal nature of economy banks on *an abundant past* rather than on *an open future.* The drawing on a credit that is necessary to make a promise is an investment in the immanence of time. Making a promise demands that one listens to the beauty in noise.[46] This is indeed a high price: whoever commits to it puts the comfort of one's soul, one's intellectual sanity at risk. It is an investment to the gigantic ("galactic") *vieillesse* (agedness) shared always already by all things in the world: "here we all are," Serres exclaims, "almost as old as the Earth."[47]

> But my brain, to only talk about that, is composed of ancient parts in the reptilian manner, of other parts as new as those developed by chimpanzees and bonobos, lastly others still, incomparably more recent. Layer by layer, it could be dated like those cliffs whose different strata sink more and more deeply into the past. Likewise, my DNA appeared, of course, with the union of my parents, who built it the way cards are shuffled, but in its own structure it is more than three billion years old; even older still, the atoms composing it and me go back to the fabrication of hydrogen and carbon by the galactic energy of the Universe.[48]

When Serres speaks of a "real age" that is common (in the sense of "same") to, "old men and the newly borne, grandchildren and grandmothers, animals and plants, friends and enemies, to all that are carrying DNA," he is showing us a model of how to hold on to the idea of generational descent in terms of *dia-sequentiality*, that is, without submitting to the linear branching of the tree as a law of *con-sequentiality.* Before time, all is equal at any instant, as he says, an equality that is contracted, "in two fractions: one minimal, the individual age, the other much larger." Universal belonging. Serres's *La Grande Vieillesse* not only introduces *a novel notion of equality,* and hence the possibility of a justice that no person or people can avail over (hence his notion of freedom is a metaphysical notion, where one is never entirely "free" but at least serves no one in particular). It also allows Serres to complement Darwinian evolution of speciation and differentiation with an inverse of de-speciation and de-differentiation. He calls this Exo-Darwinism, and it is crucial for his rejection of seeing the natural and the artificial as either categorically or dialectically opposed. If Serres's doctrine of an atomist materialism of identity had to be reduced to one vector, it would surely be the search for an arrangement in which the hard and the soft, matter and code (as logicial), share a common principle and origin.

Quantum Writing: Substitutes Step in to Address Things Themselves

Serres's universe is not an empty container, an impure vacuum, a spatial storage memory or the sterile template matrix of a prepared data bank. It is a universe that is temporal coextensive to how it is spatial, thus this time is not properly processual (continuous): a process *presumes* a beginning and an end. Serres considers it entropic rather than processual, in-definite, trans-actional, accounting; it actively "banks accounts" of the time that passes in the transits immanent to its massive transversality. Serres's intuition is close to Henri Bergson's in *Matter and Memory* (1896), but with a crucial difference. Serres lets go of the idea that presumes one *longue durée* as an overall integral:

> The womb of a pregnant woman experiences a million biochemical reactions per second; while I write this word, my organism is producing almost as many. As from a horn of plenty, the innumerable gushes forth from the instant. Celebrated by the ancient moralists and repeated by ten parrots, this moment in which I'm speaking and which flees far from me suffices for matter, for life, for thought to create bouquets of a thousand particles, to multiply as many cells or kill them by apoptosis, to conceive metaphysical systems. The instant depends on the scale: let Gargantua sneeze and millions of Lilliputian peripeteia unfold at length on the little theatre of his splutter.
>
> We shall understand nothing about the sky, the Earth, life, lastly ourselves if we continue to refer our perceptions to the time of history, short, and to build our culture upon its brevity; likewise the present moment reveals itself to be interminable. Two changes of scale discover "two infinities, of greatness and of smallness," hiding in duration.
>
> Better, a piece of knowledge and a corresponding experience have just changed our lessons on the internal consciousness of time: our organism includes, as we now know, dozens of clocks—cardiac ones, digestive ones, neural or molecular ones— all of them disrupted by the jet lag at the end of a long flight across longitudes. How are we to think the instant and duration without referring to this internal, circulatory, existential discomfort, whose appearance indicates the organic knot where our relation to time or to the sum of durations indicated by said clocks is constructed, clocks unknown to Bergson, Husserl or Heidegger, none of whom had ever flown across the ocean? Don't these clocks play, in time, the role held, in space, by the compass and orientation? Jet lag: wild fluctuations of the chronic compasses.[49]

Universal time is the non-denumerable, the "uncountable summing up" of all real durations, and as such it is always in a state of unease—like our organic time after a long-distance flight. It is uncountable in no means other than the organization of the passage of massive time in the ageing of the universe. This organization involves not only causes and consequences but also the codes that establish both: "Things don't reduce to causes but also set down codes. Things act upon each other, of course, but also make signs to each other."[50]

Serres does not think of a generalized materiality of memory, as Bergson arguably does. He conceives of it as generic and atomist. Time does not *flow* like a river; rather, it *percolates* like one, as we have seen in Chapter 3. Objects, modelled indirectly, in their chance-boundedness, behave like channels that in turn behave like river banks, through which the streaming of time bifurcates and trickles through. Memory separates and mixes in the all-that-is-time; it dissolves and concentrates in the becoming of all that endures for a while. The time of the universe is entropic, whereas the behaving of things is negentropic activity. Things conserve traces; they "conduct" themselves like memories: "All things, in principle, behave as memories. The universe banks accounts. All things are of number, the memory of the world conserves traces."[51]

When Serres maintains that "all things are of number," he is not saying that all things are unitary "in-dividuals." They are of number because signals are signals (and not signs) insofar as their recording happens in terms of numerical *chiffres* (figures). Numbers are crucial for maintaining an asignifying notion of code. They allow reality to figure in rationality. Code can be asignifying only insofar as it is always both, alphabetical (a finite set of elements) and numerical (scalar units of infinity, angular degrees that partition a circle). In that precise sense, code is alphanumerical: A logics of the alphabet, which on its own hurries forth straight to omega (the ultimate destination, the locus of action in the principle of teleonomy) is transversed in code by the numerical figuration of the alpha. Within alphanumerical code, if one says A, one need not necessarily say B, to put it plainly. The linear transport according to order relations is always intercepted— *para-sited*—and makes room for various transactional transcriptions. In other words, the alphabetical is forced by the numerical always to accommodate and account for all the mutations that arise from these transcriptions: this is why Serres's equipollence between reality and rationality cannot, in principle, admit for any "rest."

Within communication so conceived, the ideal completion inherent to the alphabetical principle, Omega, turns into *the real material support* that sends *News genuine to the world*,[52] the support of Serres's "transcendental objective" which receives information like a plant receives light. Omega, here, is like the cipher disk of any clock at all. It encrypts the measurement of temporality in all the units of its uncountable durations. The *telos* principled by invariance (rather than teleonomy) is nothing-at-all and anything-at-all; it depends upon instruction. How? Through the multiplicitous manner in which the numerical that registers reality void of any attributed significance, counts within the alphabeticity of the accounts rendered significant by rationality. *Alphabetical writing* becomes alphanumerical *coding* (dis-ciphering), or rather, it becomes *quantum writing*.

With Serres's white metaphysics, the algebraic group of all concepts of which each encompasses lucidity like spectra do, *quantum writing* no longer conserves a gospel of a beginning and an end of time; instead, it conserves the secret key that belongs to no-one-in-particular and every-one-in-principle.[53] It holds the passwords to a delivery from the a priori guilt that demands restitution for an original debt that keeps accumulating in alphabetical writing (with every generation), and in proportion to the wealth of meaning that is guarded by a discourse's legitimate statements.

In terms of generic identity's "Real Age," all things are equal; entropic noise. But insofar as they are addressed through quantum writing that *grants* them this entropic reality (by the ways science *dates* its objects), all the things of the world have inscribed in themselves the protocols of an "attribution of value" that knows no partition key or distribution scheme taken to originate in a beyond of the obligations that contract them. Quantum writing articulates these obligations. The obligations of an equation in terms of equipollence between rationality and reality do not use up deposits, and thereby, inevitably, foreclose a future for some of its openness. They descend into durations, unlock the depth of an instant and thereby create novel deposits. The promises with which the articulations of obligations fly are propelled *by having taken, daringly so, an instant for a cornucopia.*

An equational formula in the scalar (equatorial) terms of an economy of minima and maxima *takes credit without being guaranteed, by any underwriting authority outside of oneself, to be protected in doing so.* The mastership (of which we spoke earlier) does not start with violence and result in possession; it is a mastership that starts by drawing credit at no other risk than one's own sanity. It is the risk borne by any genuine thinking, thinking that invents and augments, "harbors": the stability of mental states is a fragile thing exposed to the entropic oceans of contingent options of rationalization. The transactions of this economy are facilitated by dis-ciphering the equational formula in the probabilistic and dia-sequential (alphanumeric) terms of equipollence. Such dis-ciphering takes credit from the agedness banked in the universe, banked among all else in the objects of the world. It settles this credit as soon as the equation is articulated in a solution space wherein which *it augments the identity contracted therein* (as promised by the articulation of the equation). Obligations are in contrast, then, to debt, which has to be repaid later; obligations are settled in the very act of making the promise for which the meaning of that which the promise contracts, originates. It is settled as soon as the making of the promise dis-ciphers this contractual identity in a manner that gives a novel code of how to transcribe it, without any loss or reduction. It is settled, however, *at the cost of facing* with the novel obligation also *an increase in shadows that are cast by the lucidity diffracted between one conceptual spectrum and another.* These shadows suggest, to the gnomon, novel ways and manners of taking stock of temporality. Obligations then make novel deposits that draw from the depth of any one instant which they encrypt—and thereby make last a little longer—*as the well of durations yet to come*, durations that creatively conserve the "Real Age" of the all-of-time.

The Incandescent Paraclete: Tables of Plenty

A process of hominescence substitutes for that of our fact.
Michel Serres, Hominescence[1]

*The question of evil is being raised once again, given the responsibility of our
sciences, our technologies, our truth. What are we to do about it? Some philosophers,
including Leibniz, are lawyers by vocation; others are prosecutors, like Socrates
and our contemporaries in the social sciences, who are ready and willing for police
duty; others, finally, judge, like Kant [. . .] In Greek, Paraclete, the lawyer, carries
the name of the Holy Ghost; and in Hebrew, the prosecutor is called Satan. Can
philosophy escape this courtroom? What should be said there today, when science is
speaking to law and reason to judgment?*
Michel Serres, The Natural Contract[2]

Equatoriality Generalized

In *The Parasite,* which Serres calls "a book on the problem of evil,"[3] several systems of
communication are discussed as different models. Among them is one entitled "The
Invention of the Paraclete, on the Pentecost." Serres presents it as a system of ever-
multiplying idiolects that replace a coercive common language. There is something of
a utopian character to this model of "glossematics," because it posits as systematic a
certain kind of communication—one that counts as the exception to Serres's overall
approach to the theme of communication in *The Parasite.* The book is devoted to a
model of bifurcating communication that is at one and the same time facilitated, but
also corrupted by the interceptor, by the parasite. Communication is a unifying natural
force in this model, but one that is tempered in any number of ways and manners,
via a thousand different scales such that whoever may be at the top of one chain (of
give and take, of saying and listening) on one scale, is at the same time, in parallel,
implicated in many other positions in other chains and scales. So communication is
a given, for Serres's approach. It is where he starts out. Communication is happening;
it exists; it is taking place on many scales, in many ways. In *Genesis,*[4] a book in
which Serres aims to "introduce a new object" for philosophy, he calls this mediate

understanding of communication the communicational footprint, the *ichnography* (from *ichnos*, for "trace"), also the "querulously beautiful" (*la belle noiseuse*). The new object in question is "the multiple" as such. The multiple may perhaps have been thought of before, he claims, but it has never been *sounded*.[5] Serres's communicational footprint is an architectonic that translates between the harmonic and the geometric domains. The multiple as such, hence, is an architectonic object of communicational nature, it is what Serres calls a "quasi object"; we will elaborate on this in Chapter 7, "Sophistication and Anamnesis: Remembering an Abundant Past." Its form as an object is geometrical, and lends itself to criteria of precision; its consistency as an object is arithmetic, and lends itself to criteria of rigor. But as in quantum physics where position and extension correspond to different instrumental setups that cannot be adopted simultaneously, this new object's nature is communicational in that rigor and precision actually keep it noisy, unsettled. Either the object can be studied with precision (formally, geometrically), or it can be studied with rigor (arithmetic, calculus). I want to suggest in this chapter that it is the invention of the multiple as a new object for philosophy, capable of translating between the geometric and the harmonic, that is at stake with the system of communication that Serres calls "the invention of the Paraclete, on the Pentecost." Pentecost is an important date in the Christian liturgical year. It names the Sunday that follows seven weeks after Easter. Pentecost is held to celebrate the event of an *immediate* kind of communication, where, so it is said, the Holy Spirit, the Paraclete (whose name literally means "the one that can be called to one's side," from the Greek *parakletos*) was being "poured out," descending to earth, speaking in tongues that spread and disperse. Tongues seemingly of fire; the tongues of many strange and unheard languages. We can think of Serres's exceptional Paraclete system of communication in just such terms: as tongues as if made of fire, tongues spoken within the ideational domain of what might be called "generalized equatoriality."

The topic of this chapter is the problem of evil and its relation to geography and reason, which is at the heart of Serres's philosophy of communication. As we will see, for Serres, it is reason's claim to hold the truth, on one particular kind of grounds or another, that aligns it with what he calls "the Martial rule of hatred."[6] The new object for philosophy is one that *is* reasonable, but always in abundantly many ways, on grounds of plenty rather than purity. Serres begins his book *Genesis* with a kind of motivic key in "A Short Tall Tale" ("Conte Court"). He writes, "I was sailing along that summer, under a dazzling sky, and drifting lazily in the winds and the sun, I found myself, one fine morning, in the green and stagnant waters of the Saragossa Sea, at a mysterious spot where thousands of tiny sparks, all shapes and all colours, were glimmering crazily in the early morning light."[7] The tiny sparks are all bottles: "there were countless little vessels, and each one no doubt bore its message," each with its weight and its small undulations "each carried its hope and its despair."[8] They have all been lured there, in this mystical body of waters in the form of a system of ocean currents circulating clockwise, from everywhere in the world, by the "coiling wind."[9] The Sargasso Sea is the only body of waters that is bounded by whirling currents rather than being contained by land masses. In the west, it is bounded by the Gulf Stream, in the north by the North Atlantic Current, in the east by the Canary Current, and in the south by the North Atlantic Equatorial Current. The one who can be called to one's side, this short story suggests, is *objective messages*—messages as objects, objects that are multiples—objects that are multiple and, *as such*, can be said to refer to the "ordinary lot of situations."[10] In the Saragossa Sea, Serres tells us, he almost foundered on the following night after the encounter with this extraordinary site of bottles. It was these bottles, each containing its message, that saved him; not because he could read their messages well but because as objects with cryptic content (messages in bottles), they lent themselves to being tied together, such as to form a float with which he finally "made his way back to Bordeaux."[11]

In this chapter I want to discuss the model of Serres's system of communication he calls "Invention of the Paraclete, on the Pentecost" from the aspect Serres actually foregrounds: as "an invention." In Serres's philosophy, it is scientific knowledge that is to speak truth:

> The multiple as such, unhewn and little unified, is not an epistemological monster, but on the contrary the ordinary lot of situations, including that of the ordinary scholar, regular knowledge, everyday work, in short, our common object. May the aforesaid scientific knowledge strip off its arrogance, its magisterial, ecclesial drapery; may it leave off its martial aggressivity, the hateful claim of always being right; let it tell the truth; let it come down, pacified, toward common knowledge. Can it still do this, now that it has vanquished temporal power and reigns in its place, a clerisy? Is there any chance of it still wanting to celebrate a betrothal between its imperial reason and popular wisdom?[12]

From relating Serres's book *Genesis* to this exceptional system of communication that we have been considering, something is being revealed. That system of communication raises questions with respect to the reason of objective insight and with respect to the status of geography for reasoning. We must relate reason to geography, Serres maintains, but the graphs of such reasoning do not depict a territory. They give instruction for the sounding and fathoming of the multiple, this new object by which scientific knowledge is to communicate itself. Serres speaks of a "system" of communication, but we must not confuse this with a general idea of systems as commonly understood. There is one statement in particular that is key to Serres's communicational physics, or natural communication. And this statement appears only circuitous and contradictory when considered from a more conventional (analytical, epistemological rather than "communicational") system-theoretical point of view: namely, that a model of the system is also the system of the model. Communication, in Serres's understanding, is physical—this is why I relate the invented Paraclete model to a real embodiment of it (nature), a corporeal reality that is incandescent. Serres's exceptional model of communication must be related to its *capital* (head, intelligence, the Paraclete) and solar *nature* (the universe with which the capital (universal intellect) is coextensive and to which it is immanent, in its physicality).

Coming of Age, Liking Sunset and Sunrise

Sowing: geographical sites aren't subject to simple reasons.

Michel Serres, *Rome*[13]

The new meaning spread everywhere starting from wind and noise. Not a single language translated in several languages, but several spoken and several heard at the same time.

Michel Serres, *The Parasite*[14]

"The modern West [*L'Occident moderne*] benefited from a rare bit of luck," Serres maintains in *The Incandescent*.[15] What is it that has favored the Occident? Is this claim no more than another testimonial to the superiority of the West over the East, to the spirit that sets out to do intellectual work in the morning, and that settles to reflect on its own capacity in the evening after having built a home for its reason, only to wake up the next day and begin again, ambitious for greater perfection?[16] Yes and no. As Serres puts it, "geographical sites aren't subject to simple reasons."[17] If Serres evokes, like Plato and Hegel, a heliotrope in his preoccupation with the sun, this figure nevertheless does not belong to the register of immaterial fancy, where it would ultimately prove inconsequential because the sobering dawn of the next morning would impassively reassert itself as the only certainty. But a key element of Serres's thinking is a very strong notion of the object that, in a certain way, goes countercurrent to the steady passing of time. Serres equips his heliotrope with the mechanical theory of a *generalized equatoriality*, a a-territorial equatoriality that lends itself retroactively to support *any* geopolitical argument whatsoever, and this means: no such argument at all. This is the main topic in this chapter. What then is this rare chance Serres is talking about (in his use of a heliotrope) from which the Occident has so benefited?[18]

The first book of Serres's foundational triad is entitled *Rome*, the second one *Statues*, and the third one *Origins of Geometry*. This appears to support a certain reading that Serres's engagement with a science of history is, ultimately, all about continuing with a science entirely within the Catholic legacy. But does it really? Serres begins *Rome* with a fable entitled "The Greatness of the Romans: The Fable of the Termites." It tells the story of Rome's foundation, in terms of a simple algorithm, that can be read as a materialist version of René Girard's mimetic theory of a mechanism that accounts for the fact (not the fashion) of social orders. According to this theory desire is the motive force, but one that is itself motivated by what the others desire. Like knowledge and love, as Girard is ready to acknowledge, desire is insatiable (and therefore also irrational), because it flowers and prospers from a sort of vicious circle—that kind of cycle in which action and reaction intensify one another. In Serres's fable, he too sets out an intensifying kind of cycle. His materialist version goes as follows: if you have something to give or contribute (the termite finding a ball of clay), you deposit it where the others are depositing it. What is attractive and gets attention, gets more attractive and gets more attention. Why pursue such a simple approach to a theme so weighty as that of the foundations of knowledge (the theme of the book series) and that of the foundation of Rome in particular (the first book of the series)? There is much in philosophy that can be viewed as a search for how to break out of such vicious circularity, by way of concepts that can be purely and entirely *formally* reasoned. Such reasoning too would be mechanical, but it couldn't be circuitous, and it couldn't be automatic; otherwise it would not provide the means necessary to keep its "formal" concepts "pure" (unambiguous identification of causes and consequences). According to this tradition, philosophy seeks a language that is not metaphorical, a language of concepts in which sense requires no handling in an economical manner. Against this philosophical tradition, Serres takes a similarly uncomfortable stance as Girard does with regard to many social theory approaches and also some branches of theology; by looking to find a *mechanism* that produces a particular phenomenon quasi-automatically rather than looking out for "purely" reasoned concepts that could claim to explain a phenomenon exhaustively and consistently (in this case, the phenomenon is that of violence). Girard maintains that there is something sacred about violence and that no culture can rid itself of its power. For Serres, Girard has thereby "naturalized" the sacred to a certain extent. And Serres does the same with regard to wealth and power and their acquisition by one city or another. "The ancient city was for a long time a concept; it was a type and even an ideal; it's time to see it simply as a set [*ensemble*],"[19] he explains as being his ambition for the book. There is a certain prescription to reason, he maintains, and this prescription dictates a simple algorithmic rule, the martial rule of hatred: "Reason is not a law imposing itself on the illegal; it is not an order to which disorder must submit; that reason is pure hatred. Its true name is hatred and its final production is the monstrous, glaring bright god of hatred. Pure reason, pure hatred."[20] We should think about reason in terms of prescription, not as a law of justice in any pure, immediate sense. The tables of the givens, so far as reason and its prescriptions are concerned, are set in plenty: we should think about reason as abundant, rather than sufficient or pure.

> Justice is founded on prescription. After a period of time, no one will hunt the guilty parties down any more. The sons of the guilty are not guilty. The sons of the victims are not victims. Morality is founded on prescription. What hell would a life be without forgetting or pardon? Rome is founded on a murder, and this murder refers to another murder. He who digs the foundation finds a head at the Capitol, a body, a skeleton, a mass grave. He finds Cacus beneath Remus. And so on. He who enters into a foundation enters a tomb; Rome is the city of tombs.[21]

Let us come back to our initial concern with Serres's valorizing of the Occident. For the modern Occident he is writing about, Serres maintains, "the decisive events for salvation took place elsewhere than in its home and concern other personages than the gods of its culture."[22] The grandfathers of Occidental culture themselves—be it Anaximander, Thales, Homer, Socrates, Plato, Aristotle, or Aristophanes—cannot count as the personae who engendered it. They cannot occupy central places in the narrative of the plot Serres begins to unfold here. They cannot be the authors of a modern great story (Serres's *Grand Récit* if it were read as a continuation of the legacy of the Catholic institution) that is to resume the scientific knowledge of our time;—the

Occident Serres is writing about is precisely *not* a land of (the) father(s). When he elects to begin his triad with Rome, it is because Rome was pitched first as a military camp and figures, legendarily, as the city founded upon murder: "Rome is the city of tombs."[23] This is the departure point for his treatise. It is not, however, what he tries to *account for*—Serres adopts this gesture, as he emphasizes repeatedly, from Livy, who entitled his book on Rome *Ab urbe condita. Starting from* the founded city, but *in a distance* to it:

> *Ab urbe condita.* I admire Livy's title; I wish to leave it untranslated. What is said here is the foundation of the city and designates the book that follows the foundation. Yet the city is never completely founded; the thing is never assured. It's the same for us, I mean for knowledge. Everything said here is said at a distance from the founded city; everything only has existence through this distance, through the length of this separation. The essential thing is the *ab*, or the *from*, which are, in fact, *a starting from.* A reference point, a point of departure, a bursting place.[24]

Those who try to explain the foundation of a city will need to explain union—but union is what cannot be explained, what must instead be treasured. Livy, by Serres's reading, *reports* tradition. He accumulates all the words for union and treasures them in his account. He does not "pay" anything back; he does not try to sort things out and give an explanation. He writes a statuesque text that remembers Rome as the city founded on death—a text that could claim to explain the foundations (death) of its subject (Rome) only in a statuesque manner. As such, it is a text that keeps the secret of Rome (namely, union as that which cannot be defined). This secret (union) is not a secret that could be conceived, either clearly or enigmatically. Most of all not *purely.* As Serres puts it,

> *Ab urbe condita.* Foundation is condition. Condition is union: what's placed, put together, what's put in stock, put in reserve, as in a safe place, densely packed, therefore what's hidden, sheltered from sight, from knowledge. … *Ab urbe condita,* from the union of the city. This condition is buried, is hidden. This union is hidden. Yes, we don't know what union is. We thus never cease founding the city, founding Rome here. All these words of concord and conspiracy, of contagion and counsel, are terms of condition and infinitely vary the very act of founding. We vary from not knowing.[25]

So what does Serres do then, in his first book on the foundations of knowledge? He treats legendary accounts. He treats the legends as if they were objects that can be touched, that can be handled, that one can get familiar with *without* identifying with them. Rome is the city founded upon tombs, he emphasizes. The reason that animated his book on Rome (arguably, any of Serres's many books), is one that "dreams of reconciling Democritus and Newton, of weaving together stochastic sowing and the law of harmony, dark chance and clear necessity, order and dissemination."[26] What can one learn from this Occidental legacy then if it is not to "properly understand" its power of unification? If one acknowledges, as Serres does, that this power, in the case of Rome, followed objective physical laws and was merely benefitting from fortuitous circumstances? What is to be learnt from graveyards, he claims, is how to engage with foundations that not only manifest the necessary but also contain the possible—foundations that are the work of an anonymous multiplicity of people, a massive kind of uncoordinated action that produces a masterwork without having had a plan for it. Not by algorithm alone, but by algorithmic rules together with actively treasured and sustained memory. These are just the characteristics Serres attributes to "noisy beauty," a term introduced in *Genesis*, published one year earlier than *Rome*. He calls this noisy beauty "the ichnography," an architectural term for the geometric footprint of a building. It is this same beauty that his entire philosophy of foundation is committed to, in how it is to facilitate building, conditioning, founding.

> Imagine the ground of Rome after a millennium of trampling by the Romans. Imagine the earth of the forum after the pounding of the feet of the mob. And now, decipher that ichnography. This is the final painting of the Herculean meadow, this is the initial painting of Rome; these paintings have prescribed every direction or meaning. There is prescription of every direction or meaning before the inscription of a single direction or meaning. In the beginning is the ichnography. That is to say, the integral, that is to say, the sum, the summary, the totality, the stock, the well, the set of meanings or directions. The possible, capacity. Each defined direction or meaning is only a scenography, that is to say, a profile seen from a certain site.[27]

Hence, it is not the grandparents of the Occident, or even their gods, that must count as its authors. "The modern Occident has benefited from a rare chance," Serres maintains. And further: "In no longer deifying their own soil, Western peoples thus separated the spiritual from their native roots. Thus they left the pagus of paganism, the little gods of their fields. The West has carried out said deterritorialization from its origin. Or rather: it owes its sudden bifurcation to this coming unstuck with respect to the soil."[28] And indeed, he refers to the forebears of the West not as natives to a soil: "neither the holy geography nor the holy history unfolded beneath their feet, these persons do not speak their own language."[29] For Serres, as we saw earlier, saints do not impersonate the sacred. They attend; they tend, to it. They also challenge it, but not with the intent to represent it legitimately.

> When a father doesn't beget, when a son only knows, here below, his adoptive father and when a mother remains virgin, the familial schema breaks in addition with the ties of blood. *Hominescence* describes the Holy Family, freed in this way from this chain. We have no definition; we are neither born from a defined place nor from a defined lineage.[30]

They choose to live their beliefs vulnerably; this is the affirmation of a person who does not seek to identify with the sacred, to claim any of the vicarious positions in a sacred order *personally*, justifiably, powerfully. There is spirituality to intellection, but not a disembodied one. Serres acknowledges the presence of a holy spirit, embodied in all things natural—and seeks to model it in a system of communication where he speaks of an *invented* Paraclete, without claiming that his model would identify it. This invented Paraclete in his model of communication as a system is simply the placeholder for the position that "connects many to many without an intermediate."[31] We will develop this further shortly.

It is crucial to keep a *holy* spirit apart from a *pure* spirit, which would be a sacred spirit. Serres maintains that "such a brusque bifurcation," in casting off from chthonic accountability by affirming spirit as holy rather than as sacred, is "its [the modern Occident's] condition of possibility." Casting off from chthonic accountability, by merely attending to, tending to (inventively even) a spirit that counts as holy, this gives an eminent role to a thinking in terms of a certain convertibility of "truth." Serres posits this "exceptional system of communication," in *The Parasite*, as one of economical communication. Could this be its eminent role? In Rome, Serres maintains: "economy is the continuation of war by other means."[32] Economy is not innocent, certainly, but it begins to complicate the martial rules of hatred and pure reason. It considers the relation between a thing and its value as negotiable according to circumstances. This means that we have to rethink the question of the price of truth as well as that of the priceless, as recently posed by Marcel Hénaff (we return to this in the beginning of Chapter 7).[33]

For now, we shall leave aside Hénaff's question to ask instead: Can we speak about this in the voice of the neutral "it," the "one" that could be anybody's voice? Can we account of it in terms of data? But accounting to whom, if not to some other chthonic soil on which this thinking itself (that of an economical, exceptional system of natural communication) has prospered in the past? This, as I understand it, is the motivation for addressing the earth, the universe, in terms of its agedness (data). It is the positive goal behind Serres formulation: "a process of hominescence substitutes for that of our fact."[34] This process of hominescence, like the agedness of the universe according to his metaphysics, must be considered in *material*, not in purely *conceptual*, terms. Serres adopts the gesture he admires in Livy, to set out *ab urbe condita*, to acknowledge a process that substitutes for a fact. He writes:

> As one thinks of luminescence or incandescence to accrue or to dwindle, by openings and occultation, a light whose intensity hides and shows itself in tremulous beginnings albeit continuously ready to switch off, and just as adolescence or senescence advances towards age or senility by regressing, both of them, to the involutions of an infancy or to a life which they, regretting, leave behind quickly. As efflorescence or effervescence design processes that are marked by this particular kind of ending called "inchoative," an adjective that designates a start, here of bloom, of surge or of emotion and as an arborescent plant acquires, step for step, a ramified form, the port or appearance of a tree—a process of hominescence has taken the place of our fact, but without knowing yet which human it will produce, magnify or assassinate. But have we ever known?[35]

Material, and not purely conceptual, terms—this does not imply without reason, or without intelligence. But it means relinquishing any possible claim for innocence as a point of departure. Goodness is something to strive for, certainly. The intellectual spirit is holy. But there is no referential order that could give us a pure idea of how to recognize it: a pre-Christian meaning of "holy" is not possible to determine, so the ichnography of etymological dictionaries cautiously informs us. But probably, this ichnography suggests, it was something like that which "must be preserved whole or intact, that cannot be transgressed or violated," and connected with the Old English *hal* (related to "health") and the Old High German *heil*, "health, happiness, good luck." *Rome* is the first book of foundations because, for Serres, the cultural conditions in Rome apply generally for the whole Occident he is writing about. His idea of "Occident" is not a territorial one; rather, it comes to be the index of non-territoriality within what I have called a "generalized equatoriality." In this precise sense, the Occident is wherever "the decisive events for salvation [*le salut*] take place *elsewhere than at home* [emphasis added] and they concern other people than the forefathers of their own culture."[36]

This is what is at stake for quantum literacy; the capacity to recognize this state of affairs. It is the acknowledgement of an intra-material logicial at the heart of quantum literacy which allows this perspective. Let us recapitulate. Economy that banks quanta of the universes' agedness; reasoning that proceeds by drawing credit from the enormous source of light's speed when addressing the world in its agedness; a philosophy which claims to reconcile these factors in their dual identity as universal and as singular—these are some of the formulations we developed so far. What is their importance? How do these formulations serve to find—*what is their import* with regard to finding—a voice capable of speaking, once "our fact" (that we are human, that we know things, that we can plan to do things, as well as arrange for things to happen more or less expectedly, orderly) is to be substituted by "a process of hominescence," an *inchoative* process of becoming human? To say "inchoative" is to speak of a process that is self-referential, circular, and yet *of directionality*, rather than a linear one whose direction is determined according to an object entirely other than itself, a process that would simply *be directed*. This voice must be impersonal, we saw, and it must be the voice in which the world speaks objectively, in which the things themselves articulate themselves. The only "language" whose words are spoken and heard in all languages is that of mathematics. For Serres, mathematics speaks physically: it captures how all existing things emit, receive, stock, and process energy and information, and produce entropy. This fivefold activity is, to Serres, a kind of operational metricity that gives *limiting functions* applicable to communicational physics. These limiting functions are thought of by Serres in a corporeal, bodily manner: Are they, literally, the five senses of the universal body-that-is-capable-of-thinking, belonging to nobody in particular and everybody in principle? In *Darwin, Bonaparte et le Samaritain*, Serres introduces these "five universal operations" (albeit with a question mark in the end) under the heading "Les cinq sens?"[37] (the five senses), referring back to his much earlier book from 1985 on embodiment and speech, entitled *The Five Senses*. He subtitled this book "a philosophy of mingled bodies"; this is to reflect a treatment of the role of articulation in speech and in things (their organization, their embodiment), with the stamp of an atomist philosophy of experience, where the supposed elements are not pure, but traces (*ichnos*) of mixed bodies. The limiting functions of Serres's communicational physics (the five senses, the five operations) must also be understood to offer a *syntactical* understanding of the *grammaticality* of atomism: the universal science that such an understanding of communication is to help prosper (in which coding substitutes for alphabetical writing and introduces alphanumerical [transactional] quantum writing, as we saw in Chapter 5), is not with pretensions to governing over all aspects historical and vague, in economical and calculative mode, disrespectful (a fault that could arguably be leveled against the early ambition of cybernetics as universal science).[38] The objective positivity of Serres's universal science is that of an *ichnography*, a geometrical footprint:

> New, surely, the universal science of today could call itself ichnography, since the so-called hard sciences interpret impressions, inscriptions, left behind by the events of the inert and the alive worlds—*ichnos*, Greek for "a trace, a foot step"— like philology, history, in short the so-called soft sciences interpret remains and the documents of human writing. This term, ichnography, designated in the eighteenth century the integral of all profiles under which one can look at an object.[39]

If there can be a science of history that is equally to count (by maintaining an equipollence between the two) as a history of science, it is because history begins with writing. Serres acknowledges this. But writing must now be considered in fashioned and fashionable terms, cast off from any identification with a pure language. Writing must be considered through codes, not as derivative to one particular alphabet or another:

> History begins with writing. Like us, all things of the world leave traces on things and on us. This assertion acquires, we have understood now, a universal importance. Just as, to understand an ordinary written message, it is enough to learn the form and the associations of characters [*lettres*] left … by stylus and ink on paper, the sciences discover, little by little, the codes beneath which lie [*gisent*] a thousand and one meanings, hidden and revealed by the glow of the universe, by radioactivity, by the climate, by fossils.[40]

Galileo said that "the world was written in the language of mathematics," but, Serres maintains, "it would be more apt to speak of code." With this transcription, he gives us the bridges to his arguments in *The Birth of Physics* (1977) and in *The Natural Contract* (1990):

> this equation of the second degree codes the falling of atoms [*la chute des corps*] as if a contract written for a sale or for a marriage. Juridical or physical, all laws are, by the same entitlement, codes. Written, the Universe, hidden behind it or protected by it, lies beneath a thousand codes that a thousand things decipher and that we decipher just as we do with our messages that we exchange in a thousand human languages. Dense with sense, the world enters history.[41]

Serres's construction of a corporeal, bodily and universal consciousness asks perhaps most of all a properly *metaphysical* question: "Is there a marked door [*une porte empreinte*], this mysterious chôra by which Plato evoked in the *Timeaus* the character of the universal, by comparing it to the female uterus?" Does science, if it thinks of itself as an ichnography, ask us to think, to *conceive*, as a universal support, the space-time for which Serres prepares himself ["*que je m'apprête*"] "to describe it, this *tabula rasa* on which all things, and all life forms on earth, all human languages leave their trace"?[42] To Serres, the reference to this question is to signify as broadly as that for which the term "nature" substitutes. This is why his importation of information theory into philosophy must be a philosophy of nature. This is why he refuses to make a categorical distinction, with his idea of the natural contract, between physical and juridical law.

But it is also why I propose to speak, in this chapter, not only of the invented but also of the *incandescent* Paraclete, as the body in which this spirit is corporeal and natural. Serres affirms that science eradicates language, as we have discussed earlier, but we also saw that he introduces a novel manner of addressing the subject positions in such a generalized sense of coded writing: the positionality of an author who determines a subject matter *reciprocally* yet *objectively* by casting off from its native culture toward a universal culture, suggests an author that is being determined, in turn, by the subject he conceives of objectively. The object determines the subject here by exposure, namely, by nothing more than making a lofty promise; a promise the object alone, autonomously, cannot guarantee nor see through to the end.

To affirm that science eradicates language, that alphabetical writing is being substituted by code-relative writing, does not mean that sheer objectivity would be produced by the novel quantum literacy. There is still a crucial subject position that is required to recognize the object as what it is. Any occupier of this subject position must bear witness to the experience of the object. Bearing witness: this means quantifying what is unexpected and irregular about the object—especially its very constitutive regularity itself must be recognized as highly extraordinary. The subject is subject in that it draws credit without there being an underwriting to grant for it; it needs to play daringly by bearing witness to the object in terms of its strangeness. In turn, such strangeness witnessed objectively, and preserved in code-sculptured articulations of statuesque, silent words can ensure no more than a finite, tempered and temporal, appearance to the subject: the object determines the subject, and the subject determines the object. All of this takes place within a theoretical domain of the quasi, the *as if*, the projective staging of how time can be counted within all of our "equatorial scalarities" according to which these objects can be *dated*, dated as in attributed a date of "birth." The idea of the Occident with which we started this chapter refers, for this reason, to a "modern Occident." *Modern* literally means "now existing, of or pertaining to present or recent times." It comes from the Latin *modo*, "just now, in a (certain) manner," and it derives from an adverb, that particular qualification of a verb that is supposed to actualize "manner, measure." It comes from the ablative case, that is, the indirect case that in some languages marks an object as "being carried away."[43] *Modern*, I want to suggest here, means that which is carried away from its origin by "the equation to the second degree," which Serres invokes when taking up Galileo's legacy, according to which nature is written in the language of mathematics. Objects, he maintains, incorporate some of the universe's excitable glow (incandescence). They are black boxes that contain some portion of one of light's immeasurable sources that got carried away with the ageing of the universe. Objects are crypts, statues that index "here lies." Thus, the modern Occident of which Serres speaks is one which seeks to familiarize itself with every location where the sun rises. But it censures identification with a culture that claims to be rooted in the soil of the land where the sun will, ultimately, set at the end of time.

In order to become universal, things need to be subjective. But being subject, in Serres's philosophy, comes about not by accounting for differences but by seeking identity in the domain of genuine unlikeliness (improbability) governed by objective chance. Seeking identity through learning how to take measure and how to reason geometrically—especially with regard to a physics that must be acknowledged as quantum-mechanical and communicational.

> When, in the midst of local violences, difference as dogma collapses, and relativism arrives at the emptiness of nihilism through the generalisation of regional conflicts, measure and the reason that demonstrates it remain, invariant and strong.
>
> They unite without opposing, assemble us without organising into a hierarchy, teach that men, whether solitary or in groups, are not the measure of all things. The metric of a new land, different from all the places listed or named up till then, is objectively imposed on that former reference, exclusively human, whose relative and contradictory rule used to rule.
>
> What idealist arrogance in truth it is to think that we invent everything, according to the colour of our skin, the twistings of our tongues, and the gesticulations of our institutions! No, we're constrained to obey something other than ourselves, to obey an obligation that our measures don't dictate, inform, or show, to obey a demonstrated metric, a new universe, completely different from all our differences. What a blow to collective and cultural narcissisms!
>
> Thinker of difference, from loving to measure yourself you delight in perennial war and domination: you neglect geometry![44]

Freedom to Serres is a metaphysical notion (entailing that one can never be entirely free), and mankind is not the measure of all things. While fully appreciative of sophistication, Serres's philosophy is not sophistry. Thales's reasoning at the foot of the pyramid is not sophistry. It is entirely objective, and yet it depends upon Thales's (subjective) existential investment, an investment to cast off, cutting loose his roots in the native soil and blood: vis-à-vis the awe-inspiring immensity of the pyramid, he invents a procedure for taking measure. He invents how to take measure in a manner that can be demonstrated objectively, on the abstract stage of *homothesis*; this is when "the theatre of measurement closes its doors."[45] This is when measurement becomes transcendentally objective, when it begins to qualify the originality of geometry. We are "constrained to obey something other than ourselves, to obey an obligation that our measures don't dictate, inform, or show"—we are "constrained to obey a demonstrated metric," and this means, in each case, "a new universe" and a new world.

This chapter speaks of the incandescent Paraclete as a way to address such obligation. Every metric, in coded writing—in writing where the reference relation stated by its constitutive equation, that between sign and signification, is raised to the second degree—manifests one way of respecting such an obligation. To raise an equation to the second degree means that the terms that feature in it can be raised to their exponents: they can be multiplied by themselves. This is the art of algebra; an exponent term in an equation can be factorized in an ultimately indefinite number of ways (it all depends upon the numerical domains admitted to play). Consequently, Serres can maintain that "the process of our fact" (human history in terms of alphabetical writing) is being substituted by "the process of hominescence" (universal history in terms of code-relative, i.e., alphanumerical writing). Universal history is history after science eradicates language. This is a catastrophe, without doubt. But not as a foreshadowing of an apocalypse, literally meaning an ultimate kind of revelation that would conclude the passing of time. Quite the opposite, as we have seen; it is catastrophe that conducts the passing of time (rather than concluding it). In a lecture entitled "Les nouvelles technologies: révolution culturelle et cognitive" (2007), Serres elaborated on the nature of this catastrophe: "it sentences us to become intelligent, and inventive," he maintains.[46]

I would like to conclude this section with an analogy that relates the novel concern with time in *Darwin, Bonaparte et le Samaritain* (2016) with the much earlier book, *The Five Senses* (1985). Here, Serres likens speech to "anointing" a corpse, in order to emphasize the embodied reality of speech in an interesting way that can shine light also for the philosophy of history proposed in *Darwin, Bonaparte et le Samaritain*. The analogy is this: the intelligence and inventiveness of which Serres speaks depends upon quantizing and measuring the universe's agedness in terms of units of light speed, *tempered* and *anointed* in the code-relative terms of quantum physics. This is to suggest that we can think of the code-relativity in quantum literacy as if with code we are dealing with a kind of balm that keeps together what will, inevitably, disintegrate and dissolve. From this point of view, the novel literacy both is and is not really new. Also, alphabetical writing depended on counting, measuring, and characterizing something ungraspable, namely, the articulation of breath. The novel literacy is no more nor less mysterious, ultimately, than the one in which we have thus far learned to think and to express knowledge. Nevertheless, dating objects is, literally so, *extra-ordinary*, from the Latin *enormis*, for "out of rule, irregular, shapeless, extraordinary, very large." This is because science deals with time spans in which nothing can, de facto, share a common, simply given sense. All things might share common factors among each other (a common root between genetics and physics, say), but the scales on which to balance those common factors with any sense of intuition capable of a single calibration (e.g., with regard to the events contemporary science dates objectively, like the appearance of life on earth, the extinction of species, or the expected death of the sun) are so distant that they feel meaningless. Nevertheless, this is exactly what is at stake in the novel subject position; it is precisely the credit borrowed on the strength of the natural phenomena. These scales might feel meaningless for an individual cogito, but philosophically, they feel formidable for the reverse reason; they are overwhelmingly, abundantly full of meaning. Quantum writing is a literacy because it brings forth a novel subject position, a position to bear witness, to cultivate the remaining undecided, to be active without arrogating judgements as the reason for our actions—simply in helping pass on what occurs (augmentation). The subject position in quantum literacy is one that helps a subject, paradoxically, *become one through ceasing to be one—* through attending to new meaning that "spreads everywhere starting from wind and noise,"[47] and through seeking to invent a metric that can be demonstrated, a geometric metric where our accustomed measures do not dictate everything. This literacy draws from a self-referential mentality, a mentality that becomes itself more fully, the more it contrives and achieves to neutralize itself.

How to Combine Precision with Finesse, or: Euphoria Contained by Instruments That Behave Like Cornucopia

But what, so one might ask, would such a non-territorial geopolitical imaginary possibly lend itself for, in philosophy? Does it not facilitate merely a safeguard kind of critique, is not merely one more brand of moralism rooting in a common sense of humiliation rather than a common sense that would lend itself to constitute a present optimistically, joyfully? Where lies its beauty, what does the ideation of such a geo-corporeal common sense have to offer? Serres's interest in the ancient meteorological tradition is not merely one of preserving historical richness, we argued in the foreword and the introduction to this book, but one of reconnecting with it in our contemporary age of the Anthropocene. His interest is in an anthropocenic meteora, where our political actions can take into account that the presence of humans on earth has turned into a natural force that manifests on geological scales. The outlook of such a non-territorial "geopolitical" imaginary is nothing less than that of a politics founded upon the ideal reference of an entirely sustainable technology that knows how to recycle without producing harmful waste. This reference is ideal, obviously; but what other kind of reference might there be at all for ideation? The crucial point would be that it cannot manifest as merely one more utopian horizon, which would inevitably be another hegemonic strategy of reproducing more of one particular kind of sameness. Its beauty lies in pointing to a kind of self-engendered geopolitical "continentality" whose grounds are not unreasonable, but also not entirely rational in the sense of a *ratio*, a finite resource that is essentially scarce. Reason must be thought of as abundant rather that pure or sufficient, we have maintained earlier. Can we imagine a kind of geopolitical continentality that is non-territorial in the sense of such an abundant reason? Let us look at one particular notion of instrumentality that might help us imagine this.

In antiquity, people attributed to the instruments used by what they called the wisdom of the weather (*meteora*) a self-referential mentality. Those instruments, for example, the gnomon, were considered automatic (as we have seen in Chapter 3). The ancients coined a poetic name for such instrumentality: a horn of plenty, or *cornucopia* (a self-replenishing container for the abundant). The cornucopia is the voluminous "figuration" that incorporated, fantastically, of course, the partition of a circle's self-referential scalarity. Counting time, as we saw, made possible by the invention of the scale of angular degrees, which draws the path of a spiraling staircase as if it were a Jacob's ladder between the heavens and earth. The cornucopia incorporates a fantastical scalarity like that which supposedly connects a circle's scale (of 360 degrees) roughly with that of a year (approximately 360 days). The cornucopia incorporates a thinking like that; no analogy, no proportion could be thought demonstrably, rigorously, without taking credit from the bottomless *reservoir* of a reason that is *euphoric* in such cornucopian manner. Incorporating a scalarity between what is genuinely unlikely, providing a path where there is none, like Jacob's ladder, the self-referential wisdom of instruments that serve as cornucopia not only provides credit that can be taken when setting out to invent a metric that can be demonstrated; it also affords the expression of this metric in notational scales—and it in turn grows stronger from the development of such notational systems. The instrumentality of the meteora allowed us to keep track of time in calendars. A cornucopia is a mysterious container of the miraculous plenty: Is it perhaps something like *the form* that can contain *the escalation* of time that passes massively?

The cornucopia is a container that grows in its capacity to give, the more of itself it gives away. What I want to suggest is that Serres thinks of nature in cornucopian terms. His natural philosophy is one where a process of *hominescence* substitutes for the fact of human being and human thought. Every event that science dates objectively in code-relative writing is not only an event that lies in a distant past: within the massive, percolative contemporaneity that the words of mathematical language are capable of capturing, every event that is dated radiates through all of time. In other words, every dated event turns automatically into an *advent*, a beginning. Every dated occurrence is made to occur again, due to that very dating. Serres's philosophy affirms a generalization of equatorial scalarity as developed in the previous chapters. So far this aspect has been framed in terms of its rule in projective, geometric thinking. Now I want to argue that Serres recognizes in equatorial scalarity something like a structure of ideation. The crucial aspect we want to ponder now is this: How can generalizing ideation possibly produce a realist, rather than an idealist philosophy? Would such a realist generalization not imply that the miraculously self-referential instrumentality was at work in *every* kind of instrument, given only that: (a) it works objectively; (b) it is finely crafted; and (c) it works with precision? Would not such a conception of measurement break all dams that keep rationality sane, and open science up to floods of unwanted and noisy data?

Let us ask then: Fictitious as it undoubtedly is, is a cornucopia an *exact* instrument? The word "exact" comes from the Latin *exactus*, for "precise, accurate, highly finished"; the past participle adjective from *exigere*, for "demand, require, enforce," literally "to drive or force out," also "to finish, measure." In any of these senses, it appears, the cornucopia is indeed an exact instrument. But can we say that this inventive figuration of a figment (the container of a plenty) is itself an act of rigorous reasoning? Rigor literally comes from the Latin *rigorem* (nominative *rigor*), for "numbness, stiffness, hardness, firmness; roughness, rudeness," from *rigere*, "be stiff." Is the reasoning behind the cornucopia, the wisdom of the *meteora*, numb, stiff, rough, hard, or firm? By no means, no! The sundial, the gnomon itself—this upright ruler—may well be so, but it marks the position of centrality in order to give it (the center position) away for staged plays of intellectual substitutions. The reasoning that links what the gnomon indicates, that collects, recollects, sorts, and calculates what the gnomon's indexical records "recount," is a wisdom that must embrace contingency, that must give credit to more or less obscure intuitions. A wisdom that is subtle, excitable in the sense of overly perceptive. It is smooth, vague, volatile. Like the wisdom of the mechanic, it is saturated with *finesse*, we could perhaps best say, from the Middle French *finesse*, for "artifice, delicate stratagem, subtlety," from the Old French *fin*, for "subtle, delicate." Interestingly, *finesse* derives from the noun *fine,* meaning "termination, end; end of life," from the Old French *fin*, "end, limit, boundary; death; fee, payment, finance, money," from the Latin *finis*, "end." In Medieval Latin, it meant also "payment in settlement, fine or tax."

How to make sense of the relevance of finesse in matters of exactitude? These are the questions that drive all discourses on the arts, on aesthetics, on poetics as well as a kind of reasoning that is *euphoric,* visited by spirits, demons, and genius. It is a reasoning that is in touch with the infinite, reasoning that delights in outwitting some apparent stated necessity, necessity as a "constraining power of circumstances," from the Old French *nécessité*, for "need, necessity; privation, poverty; distress, torment; obligation, duty," from the Latin *necessitate*, for "compulsion, need for attention, unavoidableness, destiny." Such reasoning takes contingency readily into its accounts; it is perceptive reasoning that thrives through self-exposure to the touch of event, from the Latin *contingentem*, present participle of *contingere*, "to touch." It is reasoning that is in *contact* with the infinite—it has the confidence to recognize in-determinedness where others see destiny. It appreciates the reciprocal touching (from *tingere*) of sums that do not add up when they are subsumed under a common roof.

But what is the origin of this credit? What nourishes such confidence? The credit at stake in Serres's natural economy collects naturally—like dew from the Bermudian islands (islands that are further west than West[48]), perhaps, this rarest of all mythical liquids that descends from the heights of night and can be collected in early mornings from blossoms, leaves, and grasses. Witty reasoning, just like the wisdom of the *meteora*, does not draw its strength from any one account of the ultimate in particular, of the first or the last. It seeks to experience the now and everywhere in its *advent*, that which is to come, and it reasons in terms of *finesse*: it makes the telos, the end, perfection—the objective transcendental referent ground of its rational models, the *homothetic* spaces of geometric projection—into its ideational (but not ideal) and lofty (abstract but not unreasonable) support. It is a support that prescribes, but is also receptive to the inscriptions of such witty, angular (scalar) reasoning.

The (Mathematical) Inverse of Pantopia Is Not a Utopia: Law in the Panonymy of the Whole World

An inverse in mathematics, from which I derive this figure of thinking that comes up again and again throughout the book (that of "inversion"), is the operation that yields a *chiral complementary* total to a *given* total. Chirality, let us remember, is this property attributed to our two hands, for example, left and right, and their particular asymmetry. As an idea inspired by the solstice (the turning of direction), the operation of inversion yields "a countercurrent effect." The mathematical inverse is not the opposite in definition: to compute the inverse to some given is a necessary prelude to an experiment, because it gives a scale its axis. The mathematical inverse is what one needs so as to achieve "an absolute" in operative terms: the mechanism for an operation is grasped objectively if one can *produce* as well as *undo* a certain effect. Such absolutes in operative terms (a complementary total to a given total, a negative entropy to a positive entropy) defines the latitude for our enquiry, subject to some requisite finesse and sensibility to contingent circumstances. The mathematical inverse adds to the equatorial balance of a scale the day of the solstice, the turning point of the sun's course. It is the Archimedean point daringly placed in an outside (Serres: an *hors-là*, an indexical formulation for *out there, here*), the abstract point of view that grants scope for speculation with regard to something given. Looking for an inverse, this is the very strategy of the artistic, witty but impersonal mechanic, the scientist who engages with those contingent, temporal orders of the *meteora*—the orders spanning loftily between the cosmos at large and the earth in nearest proximity and innermost self—that is, with the richest of all rich *circonstances* (cf. "Le Logiciél Intra-matériel" in Chapter 3).

> The course of events is exactly seen from our rational points of view. All of a sudden, without warning, the noise, a noise coming from the sky, a sound like that of the wind when it blows hard. It is produced locally, in a single direction and soon it fills the space, the whole space. In an unforeseeable fashion, it passes from the local to the global. It was a noise, a sound. It was an event in a corner of the system; it penetrates, invades, and occupies the whole house. It was heard; it is seen. They saw it appear. The noise is a chance occurrence, a disorder, and the wind is a flow. What they saw was first a distribution, a dispersion, but a division as well. What they saw is also what is generally heard, like noise. Tongues. Divided tongues, or distributed. But tongues of fire. It is the fire that pushes the winds, the heat that produces the gusts of air, the fire that crackles, that produces the chance occurrence, sputtering and crackling; it is the fire of force and of clarity of energy, light, power, and information. The noise is made message before the word is made flesh. It is a noise, a sound, the tongue of fire, and the meaning of the tongue of fire. The meaning that is cloven, inclined, divided like a bolt of lightning, the illuminated meaning. Toward declination and by the flame that announces itself to the eyes and ears. It is the beginning and the transformation; and it is in such a way that systems change order so easily. A fluctuation, a noise, a spark of chance: the state of things changes states according to this correct sequence. I have changed voices, and my tongue is split; I am speaking in rational language.[49]

The "object" of this focus (the richest of all rich *circonstances*) is equivalent to Penelope's woven fabric, the shroud Penelope weaves in the daytime and unravels at night. In *Hermès I, La Communication* (1968) Serres wrote a chapter in which he developed another model for a system of communication out of the myth of Penelope, where the garment comes to stand for the coded fabric of textuality, the networked, and temporal, tempered, reality of the system of communication. We need to relate the system of Penelope now to the system of the invented Paraclete in order to gain a proper understanding of it. If we relate this richest of all rich *circonstances* to Serres's interpretation of Penelope's death garment, the garment assumes the character of the writing that remembers the nature of the universe in all its uncountably many deaths.[50] It articulates the contingency—the being in touch of different temporalities—that is substantial for the wisdom of the weather. Such garments provide communication *without* interception. This is why immediate communication, as the inventive witnessing of the Paraclete on the day of Pentecost, marks the day when Hermes, the mediating messenger and the god of the crossroads, father of Pan, dies. Let us consider how Serres makes these connections in *The Parasite*:

> If the orator [this is Nature, tongues that come from wind and noise] is heard as is, the network is decentered, even locally: there is no longer an interceptor, no longer a crossroads or intermediate; there is no longer a town; Hermes, the father of Pan, died on the Pentecost. A miracle, they say; such things don't happen. I can speak and hear from West to East; the walls come tumbling down from gusts of wind, from blares of music. I can have a relation directly to some object without an interceptor coming in between either to intercede or to forbid [*interdire*]. Is the absence of a parasite so rare? Is immediacy so miraculous? Must the word [*parole*] always be a parable, that is to say, always aside, para-? No. If it is not a miracle, can we build it?[51]

Such immediate communication cannot entirely be fathomed rationally, reasonably. It is communication that builds on ichnographic grounds, by reasoning projectively, architectonically, with geometric footprints. Even though Serres's transcendental philosophy is all about mediacy, it is a *realist* philosophy exactly in that it affirms that communication without interception, in the sense of "immediate" contact with nature, is not all that miraculous after all.[52] To him, what inspires euphoria, what guides us in finding the strength to draw credit for thinking inventively, is nature that speaks in a thousand tongues "coming from the winds and forming noise." It is miraculous, yes, and can be positivized only *projectively* in code-based, mathematical, language related to projective and instrumental, geometrical spaces of *homothesis* (cornucopia). We will attend to this closely in a moment. The formulations of this universal language do not mean *something* because they can mean *anything*. Without a cornucopia, without projectivity, without an instrumental scalarity that is capable of containing the abundant plenty of "meaning that spreads everywhere," there is only the drowning in noise. It is perhaps somewhat overstressed, but the point must be clear: According to Serres's transcendental philosophy, objects "enfold" rarity, they "collect" and "contain" it within the rational regularity they incorporate; objects bury rarity, thereby objects are tombs, black boxes. This reality of contingency, of being in touch immediately, is not that miraculous; rather it is so *ordinary* that we forget about the fact of its very presence. An instrumentality conceived in terms of the self-referential mentality, as attributed in antiquity to this fantastical geometrical volume for the big plenty, the cornucopia; this is the locus opened up by Serres's transcendental objective where subject and object codetermine each other, where the Sphinx and Oedipus affirm that their lives are placed in each other's hands, where the nature of the human is caught up in an inchoative process of hominescence as long as Oedipus and the

Sphinx, the monstrous and hybrid animal, keep talking (cf. Chapter 0, "Instead of a Conclusion: The Static Tripod").

Hermes, the mediating messenger and father of Pan, dies on the day of Pentecost when the Paraclete is invented. The Paraclete brings the euphoria, the experience of immediate communication (communication without an interceptor), of being-in-touch with the nature of the universe. It is euphoria that inspires the writing in mathematical language, underpinned by geometric intuition that counts time, measures how it passes massively and dates the agedness of the universe. The Paraclete inspires the writing that remembers an immediate kind of being-in-touch, an original contingency that is the contemporaneity of nature in the universe. Remembering immediate communication as genuine, universal nature is what counts, to Serres, as mathematical anamneses (with a plural ending). Let us try to imagine how to approach the world given such conditions, for it is an architectural manner of address: If the absence of a parasite is not a miracle, Serres asks, if not every word (*parole*) must be a parable, *then can we build it?*

Building an order in which the parasite is absent, according to Serres's system of communication (the *ichnographia*, the "graph that no one has ever seen") subjects the utopia of abstract speculation to a *panchrony*, an invocation of all durations, contained in and released from fantastical instruments: projective spaces of *homothesis, cornucopian originality*. Let us retrace the argument: such containment is never exhaustive; time percolates through the statuesque statements written in mathematical language. Rather than assuming that the time of history flows steadily, it is affirmed here that time passes massively, and that keeping an account of how it does so depends upon studying the obstacles that resist its immediate passing—in short, it depends upon studying objects that can be dated scientifically. The statuesque statements written in mathematical language afford to take into account the spiral scalarity that de facto, in the reality of passing (massively percolating) time, sediments the circle within which we seek to treasure what we know. Serres's proposal of philosophy in terms of a *Chronopedia* refers to this ideal figure of a *circle* as a natural, ideational *circonstance* that can be "remembered" in all and any of its partitionings.[53] Such philosophy thereby gains— but has to bear as well—a distance that affords criticality vis-à-vis any depiction, any dating, of the contingently calculable sum of all measurable durations. Meteora takes on a cunningly extended attribution to the sum total of all measurable durations whose total is never static. The reasoning which such cunning affords, alongside with the dating of its objects by science, constitutes what Serres calls the *Grand Récit*. The *Grand Récit* is the collection of all accounts of scientific data. It is organized by the lexicon of mathematical concepts, according to a *white metaphysics*, in the Chronopedia. Its organization is infinitely challenged to estimate rigorously the magnitude of its own extraordinary fact: that science is capable of dating its objects within a compass of an-all-of-time—in a quantitative and formally precise manner, and hence not entirely irrationally, but reasonably. Serres's philosophy of the transcendental objective is a *meteorological critique* of the precision produced by measurement—a critique devoted to a truth that resides in finesse, in knowing *how* to respect what escapes possession because it is spread everywhere and always in the physical tongues spoken by the incandescent Paraclete. In *The Five Senses*, which concludes with the celebration of the death of language, Serres asks:

> The elderly speaker of exact and correct language finds himself crushed between the monstrous growth of true algorithms, which have robbed him of his precision, and the monstrous growth of the remaining mediatized tatters, which have robbed him of his charm.
>
> Why have I written about the five senses in a language long disqualified by so many true algorithms: without biophysics, biochemistry, physiology, psycho-physiology, acoustics, optics or logic … depriving myself of the long series of experiments, formulae, models, schemas, analytical calculations? Why write about an object that is disappearing, in a language that is dying?[54]

The answer he primes at arriving is this: "The five senses, still on the verge of departure towards another adventure, a ghost of the real timidly described in a ghost of language—this is my essay. I should have liked to call it resurrection—or rebirth."[55] In the light of the fact that this last chapter is entitled "Joy," the one effect that results (as especially perhaps Spinoza knew) from acquiring higher perfection, we can understand how Serres comes to celebrate the death of language as we knew it. Quite at odds with ahistorical, scientist positivism, he invents a ritualist form of dramatization in which statuesquely articulated code facilitates a remembering that is universal. The writing we referred to as code-based is better understood as "anointed" by code. This is why I decided to speak instead of code-based writing, as writing "anointed" by code—for unction, as Serres develops it in *The Five Senses*, is "made through mixture and produces a mixture," and, "there is no mixture that does not bind."[56] The art and practice of embalming encapsulates Serres's stance toward the prior brokenness that manifests in all things symbolic. *Symbolic*, so the philologists tell us, means that things can fit together, as opposed to *diabolic*, which means that things are brought, or kept, apart. In Serres's generalized, metaphysical equatoriality, they constitute two poles that coexist like South pole and North pole. It is also why he placed the Samaritan alongside Charles Darwin and Bonaparte in the title of his book on the philosophy of history.

The Objective Mentality and Character of Instruments

On the grounds that each and any instrument opens up and "coordinates" a space, each in its own embodied degrees of abstractness, Serres gives objective mentality and character to instruments: he maintains that there is a certain mentality, varying widely in character, pertaining to every automatism (automatism in the non-pejorative, meteora tradition of geometry). We can begin to address this mentality critically with Serres's transcendental objective-ism. It manifests in the formality of code that embalms, and allows to circulate, information that must be considered *soft,* because its scale is of the entropic/negentropic order that preserves discrete quanta only relative to a locally diverse, but universally massive, background sea of noise. The physics with which Serres wants to maintain compatibility when formulating his transcendentalism is not so much classical physics nor thermodynamics, but quantum physics. This is an important point mainly because it clarifies Serres's materialism. For him, matter, or materiality, is not the "other" to immaterial form—rather it is mass that confers solidity (and which features, in a rigorous (quantifiable) sense, only in quantum physics). In this sense, forces are never neutral, but are always already coded. Mass can be addressed in the information-theoretical and electro-technical terms that calculate with relative neutralities, in terms of invariances formulated by laws of symmetry (when regarding them as laws of conservation [Emmy Noether]). Relative neutralities are factorizable and balanceable in the code-relative writing of the emerging quantum literacy. With quantum physics, we need to include mass into our basic notions, next to and irreducible to, those of time and space. This is the assumption which prompts Serres to speak of an objective sense of intuition that his proposed "logiciél intra-matériel" is meant to address (cf. Chapter 3). If philosophy "doesn't have to dominate science or become its slave or handmaiden, … it must at least maintain compatibility with it."[57] If we are to subscribe with Serres to this maxim, then philosophy too must include mass among its basic notions. For him, philosophy correspondingly needs to find a novel way of relating to light and to knowledge. In quantum physics, quantum electrodynamics is the theory describing the interaction of radiation with materials. There are some basic distinctions to bear in mind here: Light radiation involves photons, which are bosons, that is, the carriers of the electromagnetic field. Photons do not have any mass and they are not charged. In contradistinction to photons, electrons are fermions; and they are considered as material particles which are charged. White light should therefore be considered on a par with material black body radiation—this has actually been the originating point of the quantum description of light. Conceptually they form a polarity. Photons and electrons cannot be thought of as independently existing, one without the other. In this sense, light can be addressed through physics, and must count to philosophy as massive and material, not as immaterial and neutral in any absolute sense. Let us hear how Serres speaks of the role of light for philosophy:

> Philosophy loves light and has turned it into the model of excellent knowledge, especially the splash of daytime sunshine. Sparkling with truth, light is supposed to chase away the darkness of obscurantism. That is an absurd and rather counter-intuitive idea, as we all know that any candle, as weakly as it may shine, immediately pushes back the shadow of the night, while no one has ever seen darkness overcome any source of light. This ideology is terrifying because if we turn the day into the champion of knowledge, we are left with only one unique and totalitarian truth, as hard and unsubtle as the sun at high noon, the star that astrophysics has eventually confined to the minor rank of the yellow dwarf. I say no to this tyranny, no to the yellow dwarfs.[58]

This is an epistemological inversion which no longer equates light to insight in any simple and direct way: *it seeks the generalization of the natural source of light, the sun.* Knowledge turns into a question of light in the aspect of mass; that is, the amount of mass that is proper to light. And this is the crux of the trouble, because within the contemporary paradigm of mass in quantum physics, light is not the opposite of materiality. It is in a fascinating way at once continuous and yet discrete with and within materiality. In quantum physical terms, light is simply the *accountable absence* of mass.

The question at the core of the Chronopedic thinking, which we want to pick up now, is thus: How to account genuinely for the absence of something, that is, without presuming what it is that is absent, without thinking in terms of lack?

This problem is precisely the one Serres sets out in his book dedicated specifically to the philosophy of history, *Darwin, Napoleon, et Le Samaritan. Une Philosophie de l'Histoire* (2016): "Historians like to praise themselves for exercising, for practicing, for celebrating memory, even though their discipline defines itself as a series of forgettings," he writes.[59] Serres thereby replies to the kind of scandal his proposed philosophy undoubtedly often attracts, attuned as it is to "dark kernels of meaning" which are in principle never fully amenable to elucidation, which we can never entirely rid of obscurity, and which is never exposed naked to the blank lights of an immaterial principle of clarity. The scandal is that he seeks in contrast to develop an objective sense of intuition from relating the formality of math to the mythical kernels of meaning. Math and myth call each other forth. They form a polarity like white light radiation and material black body radiation do in quantum physics.

Would not our knowledge lose reason if we submitted to such an adventure? To preemptively avoid precisely this, history is commonly said to begin with writing. To Serres, this constitutes the first of those "forgettings" he speaks of in the citation just mentioned. Have peoples who did not invent writing no history at all, he asks? To a certain extent, the discipline of ethnography compensates for the forgetting of the historian, he elaborates. But in ethnography as well, there is a forgetting: we learn early on in school about the descent of Homo sapiens from Africa. But we do not learn about the splendors and antiquities of black cultures—for, as he puts it, "how could we know, they would have had to write."[60] Again, there is a discipline which in part compensates for this, and this discipline is prehistory. But it only takes into account the artifacts of human cultures. Serres's conclusion: "To put it otherwise, the human constitutes the exclusive or central subject for history, for ethnology and for prehistory, even as their references. With pathetic complaisance, we mirror ourselves, narcissuses, in these three disciplines."[61] We have seen, in the preceding discussion of the incandescent Paraclete, how crucial it is for Serres to maintain that habitual human measures cannot be the sole means to orientate thinking. Science and philosophy are obliged to listen to a metrics that is demonstrable *objectively*, he insists. And this is what he celebrates in this tremendous (but also tremendously frightening) development in which "science eradicates language."

All the sciences today are concerned with processes of dating, of keeping stock of time that passes massively. This challenges our inherited notion of reason. Reason is ill-justified in terms of its clarity as well as its purported sufficiency (when taken in absolute terms). Reason must be considered as incandescent, universal, and abundant; Serres's entire philosophy affirms this. To sum up once again: Serres's natural economy begins with capital (the sun is the ultimate capital, that is why his metaphysics is one of capital concepts, "white concepts"). It does not result in the accumulation of wealth. Economic mastership does not accumulate possession but divest itself of it, as we have seen: in all our powerful and mediate ways of communicating, we must remember the immediacy of the incandescent Paraclete, by listening to how the things of the world speak, by appreciating that meaning results "from the winds and from noise" and is spread everywhere, always.[62] Reason is abundant. Intelligence is objective and coextends with the universe itself. Instruments for rationalizing the real embody self-referential mentality and character; they draw from themselves what they cannot (yet) contain; and they grow more prosperous the more they exhaust themselves in this manner. Science eradicates language. This is affirmed by Serres because with this new conception of universal and abundant reason, science is no longer restricted to *dating* only those objects more readily considered objective, natural, like the age of the universe, that of life on earth and the appearance of life forms. This same science can also begin to date "soft" objects like cultures and civilization, artifacts, techniques, languages, narratives, and so on—objects that have hitherto been categorically separated as cultural, not natural. The history of all such dating is the *Grand Récit* to which Serres relates the process of hominescence, a humanity that is being born, and always has been born, into new worlds, ever again.

Let us try to imagine such dating more concretely. Especially, let us ask where the traditional philosophical question about the sense of human history is being extended to the *Grand Récit*, which precedes and accompanies it.[63] First, we have to observe that the *Grand Récit* is exactly this, a "recital" in the properly literary sense of a thematic plot, the action of a happening that is being picked up, told again with variation, continued through altering it, repetitive but with unexpected, surprising turns. A recital combines cyclicity with linearity. In this sense, it relates also to chaos science where aleatory bifurcations evolve into outlooks that might well take on a consistency quite inverse to what the pledge, initially, had guaranteed:

> To put it otherwise, this *Récit* has meaning only when it is told in the future tense, a path I have just taken, while remaining without end, upstream to downstream. This absence of finality is the basic instrument of the polymaths [*savants*]: if it existed, they would not be doing science.[64]

We can easily see again here the great relevance (as discussed in Chapter 2) to Jacques Monod's affirmation of objective chance: invariance must be taken to precede teleonomy. This very tenet has become constitutive for Serres's thinking as a whole. Hence there is a plot to worldly knowledge, in the old poetic sense of unity of action, but this unity is mathematical and universal. And just because of this, this unity is indeterminate, but determinable. Each term in a plotted equation can be heard to speak in many tongues. The plot of this big account is a plot that unfolds in a temporal *passing* that is not, of course, in any simple way linear, but "full of unexpected, unpredictable bifurcations."[65] The equations that are being plotted, when the dating of objects is told in future anterior take on the naming power laid out in the sixfold name of the world's panonymy: from plots traced out in this way, one could not have foreseen the arrival of a particular event, and any subsequent explanation could take many forms:

> What this signifies, I want to underline again, is that when considered from above, the emergences could not have been predicted, but that, viewed from below, they

could well seem as if perfectly rational, deriving from one or several causes, or rather, conditions, that the expert discovers as necessary but which never amount to sufficiency. This logical and mathematical distinction puts the retrograde movement of truth in the light of its rigour. Before, blind; after, diverse varieties of clairvoyance. I love to say again, the *Grand Récit* may well be an account given, a novel full of suspense, tragedy or comedy, traversed by messengers delivering news, who, in making time bifurcate, increase the tension of the wait, just as we know it from ordinary history.[66]

The movement of truth is a retrograde movement; what Serres speaks of as a *Grand Récit*. There is a plot, a narrative, but it must be drawn mechanically, printed, plotted, in automatic fashion. As far as the plot is concerned, no subjective deliberation can figure therein. This would bring in a finality that dominates and directs the inchoative movement of retrograde truth that needs rather to be witnessed, by addressing the world objectively. Thus the *Grand Récit* is a kind of a testament; it counts a heritage. But whose? Serres does not seek to escape the question: "Could the *Grand Récit*, a multiple billionaire, could it pass for the Age of the Father?"[67] Evidently, the retrograde movement of truth he commits himself to obliges us to talk about beginnings. But like all scientists and good historians alike, he discredits the creationist impulse behind the obligation: while ascending a path, one is never to talk about anything else than the base station. Scientific thought is a kind of "climbing back up towards the beginning."[68] It is in this sense that the new history, the science of history that serves (by equipollence between science and history) at the same time as the history of science, studies the agedness of the universe with a clear commitment to beginnings: it seeks to account for the strangeness of its objects; it does not seek to expose, preserve, and purify their regularities.

So is Serres's philosophy an asceticism? Is it a theosophy that seeks to direct the ambitions of science to a beyond of this world? *The Parasite* (1980), perhaps Serres's most explicit book about communication, exposes its basic relation between sender and receiver of messages in the genuine sense of the word, that is, as *news*, as the human condition in expulsion from paradise: to be a parasite literally means to eat next to (from Greek *para-* "beside" + *sitos* "food"[69]). Paradise, he maintains, is the maximal scene of parasitism: God the father who only gives and takes nothing, and all beings who only receive and have nothing to give. So no, the *Grand Récit* does not determine the age of the father. But it does not refute the Catholic legacy of universalism either: it ciphers (quantizes, alphabetizes in terms of a quantum literacy) its personification of nature in Christ, and naturalizes the persona of the savior into the inventive eventfulness of what he calls the Paraclete:

> The ancient, venerable theology of the Paraclete matches in part the anthropology of exchange. When the Holy Ghost comes, so do gifts. He is the gift-giver, *munerum datar,* and his gifts are seven in number, *septiformis munere, sacrum septenarium.* The paths of the wind are not reversible; its origin is at only one point of the compass, and the owl never goes back to that point. The gift has a source but is not a point of reception. There is no exchange. What come from it are Wisdom,

Knowledge, Intelligence, Advice, Force, Piety, the Fear of God. Eliminating from the list what is properly considered to be divine, what remains is what has the characteristics of what we call information.

Fire, from which wind comes, which comes from noise, from which gifts come, is paradoxical. It heats: *fove quid est frigidum, ignem accende*; it burns; but it cools. *Dulce refrigerium, in aeste temperies*. From this source, from this mouth, both hot and cold blow.[70]

Serres's history of science, which is—in the sense of an equipollent relation—a science of history, seeks to *cultivate* the regularities of objects like landscapes, that is, relative to localities and their weather conditions. This is what orients this universal understanding of science toward the world, here and now, in its richest of rich diversities. The concepts in mathematical language, the concepts of a Chronopedia, turn to the intensities, qualities, appearances due to both, white light's radiation as well as black body radiation. They don't need to follow the logical path of negation. Absence, in mathematical language, is framed in the terms of a cipher's zeroness (cf. Chapter 2). Absence is the absence of the source of both, negative and positive: absence can only be framed spectrally, masked, by the code that articulates a cipher. Order in terms of light's mass-*iveness* is quantum physical, information-theoretic *order* in the terms of entropy and negentropy. To generalize and abstract from the natural source of light, the sun, means that we need to stop time at one moment and then seek the mathematical inverse to this moment such that the possible passing of time unfolds a *worldly* space. A geometrical space. This is how Thales finds the space for a placing of things in terms of their projective likeness, their projective similarities (the space of *homothesis*). And then comes the crucial point; after having stopped time speculatively, accounts have to be opened up for history to be taken into stock. The focus shifts now to the need for philosophy to accommodate mass as a third among its basic concepts, next to time and space, if it is to be compatible with today's natural science (physics, quantum science). As we have seen in Chapter 2 on quantum literacy, as well as in Chapter 4, white metaphysics imposes an economic paradigm of order. Knowledge is not gratuitous; information has a price. The key insight to grasp is that money as an operative, a convertible, general equivalent, is not in principle at odds with the universality of the knowledge we seek through science.[71] This is what Serres means when he maintains that all things bank time: "All things, in principle, behave like memories. The universe banks accounts. All things are of number, the memory of the world conserves traces."[72] Geometrical measurement of things *hypothecates*, takes into account "a deposit, pledge, mortgage; from *hypo-*, "beneath, under" + *tithenai*, "to put, to place." Weighting the value of things as objects that are dated sets the price for the worth at which they are being conserved in the Chronopedic accounts.

Every account of the universe's agedness, every scientifically dated object, is a pledge, a promise that takes the form of a deposit. What is being deposited is what has *not* happened but *might have*. We should recall that our obtained space for measurement is a *hypotheque* that consists in declaring that a certain duration is to count as reversible. We have seen with respect to the Chronopedia that measurement cancels out time. But it does this through *hypothecation*; it counts on what it does not know from experience. Measurement cancels out time projectively. The spaces of similarity conceived by geometry (spaces of *homothesis*) come about through generalizing the natural source of light, the sun, and such generalization is acquired at the price of maximizing contingency for a particular observation. The path from one generalized moment to another, from one dated duration to another, consists in remembering the cunning move of stopping time in order to have something to give back to the world that has been taken into account as the given, by singularizing, exposing and inflating, one of its moments. To estimate the worth of measurable information is to symbolize a metrics for such measurements that follows the scalar logic of equatorial cyclicity. Such a metrics can be symbolized by formulating a measurable phenomenon's (a positive observation's) complementary inverse (the scope of how a positive observation could be produced and "de-produced"). Put in such precise and technical terms, this sounds very abstract, but what it says in the register of information theory, is nothing other than what every mechanic considers self-evident. We speak of understanding something in terms of its mechanism if we know how to produce it (positivity) and how to undo it (negativity). Estimating the worth of measurable information depends upon a kind of modelling that represents *pass-ively*, meaning with a passivity that acts like a filter: it is passive actively in that it percolates mixtures.

The Vicarious Order of Knowledge That Is Authentic to the World

The *Grand Récit* is written through geometrical conception of the nature of the universe. One assumption is necessary for such a conception, which we have framed in terms of an objective kind of witnessing that draws credit without thereby creating a determinate debt, that accepts a certain mysteriousness without giving up rational and reasonable control. Such witnessing characterizes the subject position in the code-relative writing of mathematical language. A geometrical conception of the universe coexists with a circuitous natural economy in terms of *maxima* and *minima*, as we have seen. The assumption it entails is that the starting point of measuring is *positively*, but ultimately *unreasonably*, given (significant, but too contingent for constituting something meaningful). Our introduction of polar equatoriality into the counting of time depends on the following assumptions that cannot be reasoned: (1) there is a cyclical distribution of days and nights between two rare events when the days and nights are both are of equal length (equinoxes); and (2) there is the point when the seasonal movement of the sun's path comes to a standstill before reversing direction (solstice).

The horizon for authentic knowledge, knowledge in the sense of a geometrical conception of the nature of the universe, is imposed by the earth's mathematical scope for defining a scale: *equation-ality*, projected through *equator-iality*. Such equationality is projective rather than merely hypothetical. This projectivity is what substantiates the hypothetical, and makes it a *hypothèque*: it is a hypothesis that cannot be put forward gratuitously. Credit is required in order to substantiate its claim objectively. Obtaining information towards that substantiation is not gratuitous; it comes at a price. It is in this sense that the projection of metrical scales in terms of equatoriality demands the deposit of a kind of existential trust, a mortgage, as the very foundation on which to build.

The acceptance of, and subjection to, such mysterious riskiness can make such a mathematical scope pay out in *theoretical* observation. We can relate the notion of "horizon" to such mathematical scope that involves the deposit of a kind of existential trust: the term comes from the Greek *horizon* (*kyklos*), "bounding (circle)," from *horizein*, "bound, limit, divide, separate," from *horos*, "boundary, landmark, marking stones." We can think of it as the geometrical generalization of the singular instance of a geophysical "equator": the horizon then is the meteoric distributor of "equatoriality." Considered like this, accounts that involve the horizon, or "horizontality," must count as "geometrically symbolic"—they owe their mathematical scope to an algebraic constitution of the involved geometry. Why algebra? The mathematical form of algebra is that of formulating identity as equations; what we have tried to trace in Serres's manner of thinking about mathematics is how an *algebra of the world* could be conceived (not an *algebra of pure thought*, as Wilhelm Gottfried Leibniz, Bertrand Russell, also Gilles Deleuze and many others have looked for). How could an algebra of the world be conceived that is capable to accommodate in its identity formulations not only the mass-ive *nature* of thought, but crucially so also the percolations of time that passes massively? Combined with a metrics of the world (geometry, homothesis), algebra is the mathematical way of witnessing *theoretically* (not immediately) the kind of "worldly originality" we are looking to account for; beginnings that spread everywhere, always. The beginnings of a generalized equatoriality, summed up and distributed via the mathematical scopes that figure as meteor-ic "horizontality."

Algebra is capable of taking stock of white light by boxing something, together with its inverse, as a mechanical coefficiency of the real with the rational. In their algebraicness, the accounts of the *Grand Récit* are authentic reports by *witnesses*—and authentic means, in this affirmation of an inevitable theoretic (projective) stance, indifferent to finalities. In *Rome, The First Book of Foundation* (1983), Serres explains how he sees the relation between mathematics and myth. Algebra is mythical. It alone cannot give authentic knowledge:

> The victim is the element of substitution. The neutral element. The white element, the one that can bear every value.
>
> The black box is full of white elements.
>
> Thus myth is algebraic.
>
> I'm using "white" in the sense it has in games and in light. White is the sum of colours; it can be broken down into the spectrum of the rainbow. Thus the white token can settle on any value. You are an animal, a shepherd, a king and a god; you are white; you are good and bad, robber and robbed, murderer and murdered; you are white. You are victim, substitute or one substituted for; you are white.
>
> When you aren't white, a determination appears, a mark or a sign. Determination is negative; if you are king, here and now, you aren't an ox or a shepherd or a hero. Indetermination is positive. White is the indeterminate, the limit of the underdetermined, the whole of the positive. You are white—yes, yes, yes—you are every possible world. Leibniz had possible worlds be visited in a pyramid; the pyramid is fire, the fire is white light.

> In the beginning is the black box, ignorance, our zero of information. In the beginning is the white, every possible world. In the beginning is the victim, this relation of substitution, and this death, between us.[73]

We can see now how the magnitude in terms of which the universe's age is to be assessed, is neither an infinite magnitude nor a definite one. It is the magnitude Serres's white metaphysics organizes into transcendental bodies of thinking. It is the magnanimous magnitude that treasures a plenty, the natural magnitude of a reason that must count as rational yet abundant and unsettled. All objects that are being dated, are being *banked*. Authentic knowledge is not innocent. Universal nature, whose magnitude the *Grand Récit* recounts, is a total that originates in itself in the percolative process of taking stock of its own passing duration. It is the stock as a total whose arithmetic speaks, in the coded (and coding) tongues of the incandescent Paraclete—not in one proto-language.

> Thus authentic knowledge overflows with results and intuitions; it sets up multiple reference points grouped into constellations with forms that are as disparate as those of scholarly disciplines. Thus knowledge finds temporary truths, whose luxuriously coloured sparkle flickers and changes with the duration of the Great Story. The only lights that do not tremble emanate from planets without an original brilliance and that, as I said, behave like mirrors. Magnificent, but modest enough to be reduced to the punctual … Great in size but wavering in doubt and questioning, those truth-stars stand out against the enormous black background of non-knowledge, empty without limitations or full of yet unexplored galaxies: things still to be understood and to be grasped.[74]

What Serres can call "authentic knowledge" is knowledge authentic to the world. It is knowledge granted the right to abstain from clear lucidity, which entails nothing less than binding it to the obligation of *crediting*, on the basis of existential trust, expeditions that set out to account for gaps in the horizon, for breaks in the continuity. It means taking the trouble to deal with the notion of an indefinite temporal *total* within which time *percolates*, and relative to which no knowledge can ever be exhaustive. The concepts which grasp such accounts are capital concepts. They are concepts that draw from credit that must be taken without it being granted by something or someone: such concepts are original, or algebraically white.

Pan: The Excitable Subject of Universal Knowledge

Serres thinks about how we might address such authentic knowledge of the world by giving the third-person singular, the grammatical person for the impersonal agency at work in objective knowledge, a mythically projective face: he suggests nominating Pan, the god of all (Greek, *pan* for "all") the subject of universal knowledge (in the mathematical-mythical terms of his realist philosophy). In Greek mythology, Pan was the personification of nature, the guardian and multiplier of all things, literally "the nourisher," moved by lust and living in the woods, with a hybrid half-human half-animal body; he was pictured with horns on his head, and the legs of a goat. Pan is a god in a world of abundance, and yet he is not only moved by lust but also, strangely so, animated by desire. Strangely because in a world of abundance, desire cannot be due to lack. The nymph Syrinx, well known for her beauty and her chastity, hid from him in hollow water reeds, from which Pan invented the flute as an instrument to express his longing for her. If we address the impersonal subject of knowledge as Pan, and the world as the object that determines Pan's identity, by all the names proper to its Panonym, as Serres suggests—then we will not forget so easily that the world will always be haunted by what it desires and longs for but can never consume or own. This mythical masking also conveys that the sounds of the world's longing are sounds that are as pleasing as dreadful. Forever on the verge of triggering panic, groundless fear, contagiously transmitted and so forceful that it dominates and prevents reasonable and logical thinking: panic literally means "all that pertains to Pan" (from the suffix *-ics*). And Pan is a god in a world of abundance who is moved by lust and desire, by longing and relaxation. The story goes that he would wander peacefully through the woods, playing his flute and resting every day at noontime. If disturbed in his sleep, Pan would shout so loudly that all herds would stampede.

The world needs to rest when the sun is high in the meridian: its reference, for all the time keeping that clocks can do, is the hour of slumbering indetermination. The world, addressed in the panonym proposed by Serres, needs to be given its rightful break at noon when the light is clearest and when all appears (as if in a generalized equatoriality) with absolute regularity. Noon is no time to work but the time to remember the universe's agedness present everywhere in the world through the spectral rendering of the Great Pan. A time to remember that the forms apparently producing absolute regularity are forms that originate in death—in the death of Hermes, father of Pan, whose death coincides with the advent of Pentecost. Authentic knowledge treasured in the Chronopedia remembers the death of Hermes and refuses to think as miraculous the advent of something like immediate communication, communication without a messenger. Such a refusal informs the ambition to build a machine that can produce it. Technics, here, is proper to the world, and hence to its authentic knowledge. Yet a technical device can hardly reproduce the full present of the real experience in this world that is one of abundant reason.

> Hermes, the father of Pan, died on the Pentecost. A miracle, they say; such things don't happen. I can speak and hear from West to East; the walls come tumbling down from gusts of wind, from blares of music. I can have a relation directly to some object without an intercepter coming in between either to intercede or to forbid [*interdire*]. Is the absence of a parasite so rare? Is immediacy so miraculous? Must the word [*parole*] always be a parable, that is to say, always aside, para-? No. If it is not a miracle, can we build it?[75]

Generational Con-sequentiality

Blessed Curiosity

When Serres was invited to give a talk in celebration of the 150th anniversary of his own college, the Lycée Saint-Caprais in Agen, France, he used the occasion to share what he calls "a confession." It is a short and humorous text, full of tender memories about all sorts of more or less innocent mischief, but it also places a ruse that at once confines and upsets the honorary frame of generational sequence in which he had been invited to speak. "God has given us the endless freedom to disobey him, and this is how we can recognise him as our Father," Serres sets out, and continues, "scarcely installed in the terrestrial Paradise, Adam and Eve quickly eat the apple and pips, immediately leaving that place of delights and fleeing towards hazy horizons. Only a few months old, the infant tries to say no; those among you who raise children will learn this and know it in overabundance."[76]

The presumptuous ruse Serres has placed in this "boring preamble of mixed theology and natural history," as he calls the story of expulsion, the ruse from which he wittily distracts also by the grandness of the opening address in the first sentence, is a small change in the setup of the biblical great story: Adam and Eve give in to their human and supposedly corrupting and degenerative inclinations for *curiosities* not only lightheartedly but also naturally, in Serres's account. This mischief introduces into the narrative of the tree of life nothing short of an abundance of directions in which it might descend and branch off. What presumptuousness, indeed! One that dares to take aim, high-spirited, humorous and quick, at nothing less than the total, the ultimate sum, by unsettling the grounds in which the tree of life is rooted. Yet without, as we will see, sacrificing the morals that spring from the order that is being unsettled (paradise, corruption, and expulsion).

But how could such ground possibly be unsettled? Serres assumes that the nature of the human must, along with everything else, be thought to factor in a universal nature— a nature of the universe—whose path of descent is divine (omnipotent) and decidable (lawful) *as being undetermined* in its direction at every branching. It is a nature capable of developing in an uncountable abundance of directions, progressive as well as regressive. This nature then must count as essentially arcane, a secret that can be preserved only in a "crypt," as Serres refers to it elsewhere.[77] Of course we know the term "crypt" from the architecture of churches, but it once meant more generally a "vault, cavern," derived from the Greek verb *kryptein*, "to hide, to conceal," by nominalization of the adjective from this verb.[78] For Serres, there is a path for knowledge to access universal nature, but never a plain, pure and immediate one. All knowledge is a reduced model of universal nature, a model that does not seek to represent nature, but rather seeks to preserve and keep alive, as best as it can, that nature's character: to be secretive. The entire raison d'être of such knowledge is to serve and obey—unconditionally, absolutely—nature's secretive character. Such obedience can only be performed through disobedience, through mischief, through the comic. It can be performed by inventing a reduced model of the secret; something achievable without the assurance of being initiated to it. Universal nature's secretive character can count neither as private nor public, neither as esoteric nor established insight; rather, we can refer to it as constitutive for both in a fashion characteristic only of law, observes Serres.[79] Knowledge then embodies law in the building of a crypt, a vault, one that grows deeper and vaster, more intertwined and winding, through the fact of being frivolously explored, challenged, tested, strained in the very massive (information theoretical, quantum physical) solidity in which it is built. To keep the secret that is universal nature demands absolute obedience, yet without tolerating submission: its secret indeed has one vulnerability, namely, that obeying it can look like servility. Serres refers to the evangelist "who called Satan the Master of the world."[80] As the other to one who strives to master the universe's secret by keeping it encrypted, and who spends time in that very vault that does not cease to challenge and take issue with the earthly grounds where the iconic tree of life is rooted, the master of the world "leads you to a very high mountain, shows you all the kingdoms in all their glory and promises to give them to you on condition that you grovel before him."[81] If knowledge of the universe's nature is a crypt, knowledge of the world is the crypt's flat projection in terms that nonetheless claim the authority to represent the crypt's arcanum. Such flat projection alone can claim to produce "positive" or "negative" knowledge; the crypt in contrast embodies knowledge that is always already articulated, knowledge that presents insight only by leaving absent what it has intuited. Serres's seeker of articulate knowledge, whom he addresses as the Researcher, serves the Law. This seeker is an "official" whose duty it is to explore and challenge all the regularities that have been stated as lawful—without ever claiming to represent those regularities with official authority. "We always save ourselves by the law. Freedom comes from laws," Serres tells his audience.[82] Law binds and contracts the ambient terror of the jungle, in a manner that allows "a balance between hunting and being hunted, between eating and being eaten."[83] Law contracts violence. If those contracts are sound, whoever is subject to it can afford to live and care for all that is vulnerable as the source of all that is improbable and precious. With these elaborations, we can perhaps better appreciate the radicality of Serres's confession: "I continue to make mischief in order to bear witness in the face of the world that we are not beasts, that therefore we have left or begin to leave the hell of violence, because we are men."[84]

In Serres's humorous confession story, giving in to the human inclination to be seduced by curiosity ceases to be a tragic act. Rather, it is the researcher's official duty to enjoy masquerade, to be transgressive by engaging in the challenges that motivate desire and seduction, pleasure and satisfaction, pain and relief. This is comic, and yet it is serious: a researcher "cannot cheat."[85] For "to obey, here, consists in submitting oneself to the laws of things as such and to thereby acquire freedom, whereas cheating consists in submitting oneself to the conventional laws of men."[86] In Serres's inversive account where the universe has an active nature (rather than being imagined as either static or dynamic), cheating becomes equivalent to being obedient (to the laws of man), and disobedience comes to count as blessed rather than cursed:

> Things contain their own rules. Less conventional than the rules of men, but as necessary as the body that falls and the stars that revolve; even more, difficult to discover. We can do nothing and should do nothing without absolute obedience to these things, loyal and hard. No expertise happens without this, no invention, no authentic mastering. Our power comes from this obedience, from this human and noble weakness; all the rest falls in corruption towards the rules.[87]

For the researcher and the comedian, disobedience, as it characterizes the tragic mode of acting, is not thought to be nourished by delusions, now to produce regret, anguish, and guilt that can be relieved only by comfort derived from acknowledging the principle impotence to which such acting is always already sentenced. Quite inversely, the very possibility for disobedience comes to feature in Serres's account as that which is capable of preserving the possibility of salvation. Acts of comic disobedience replace the Scriptures as that which preserves and circulates that possibility. What in the Scriptures unfolds between the two covers of a book (or the top and the bottom of an inscription plane, be it stone, clay, papyrus, or parchment) is thereby attributed a different status by Serres: the mediacy of *the place holding* (comically disobedient) *acts* that respond to what unfolds between the covers of the Scriptures themselves—substitutes that enter the limited inscription plane or the numerous sheets contained in a book—are considered capable of capturing, conserving, and expressing a sense whose extension as meaning is *in principle* of vaster magnitude than that which the two covers (or the limiting ends of a plane) are officially entitled to contain.

In Serres's narrative, Adam and Eve taste the pleasures of transgression and disobedience without—*decidedly*—thereby falling into corruption. And suddenly there is the possibility of a distance, a genuine mediacy capable of discerning a human world as a locus in quo that spans between the traditionally purported covers of the scriptures, the "original" act of divine judgement that is said to have predicated the nature of all that is, and the act that is consequential to Eve and Adam's frivolity in tasting from the forbidden fruit, namely, the *divine sentence* with which the ancestors of humankind are sent into expulsion, the act that leaves the disobedient ones alone with the representatives of official generality as the sole gatekeepers of a source of comfort. If, however, the divine entitlement of the word between the covers is to preserve the possibility for disobedience, then the titles conveying the truth preserved within must pass on the *virtually abundant activity of possible disobedience* they are to guard in the service of duty they are to represent. Like the plane of inscription they limit, or the sheets they bind, the covers too need to be capable of capturing, conserving and expressing a sense. A sense whose extension as meaning is *in principle* of vaster magnitude than the text which the two covers are, again, *by custom of current office*, charged with preserving and entitled to contain.

In other words—and this, again, is Serres's characteristic humor—if you hold respect and esteem for official representations, then never trust them, especially while paying service to the law they represent! In an admittedly twisted but not really complicated way, as I have tried to depict, it is their entitlement as official representations to take care of their capability to compromise themselves. "To compromise" here is an important, albeit dangerous term which translates the German word *Bloßstellung*, which means something like "embarrassing exposure," a kind of personal vulnerability that comes from "lowering one's guard" (*sich Blöße geben*). The guards of an official representation would of course be the official order, so the implication is that the official representation must in turn have "capabilities of mediacy." The capability in question is the capability to transmit and pass on the *virtually abundant activity of possible disobedience* (which this representation is entitled to delimit and protect) to the official order that is charged in its turn to guard and protect those same official representations. Here we have another depiction of the self-referential mentality proper to instruments whose metrics, and hence rationality, is scalar.

Like Eve and Adam in Serres's narrative, and like the unfolding *mediacy* between two entitled limits, the entitled limits themselves must also be respected in their divine nature, and this divine nature consists in being endowed with the possibility for disobedience. This very possibility is defended in Serres's narrative; it is the factor which appropriates from the plot its function as a story of salvation, despite the frivolous masquerade of that plot's central protagonists. Serres engages as the narrator of that story's novel articulation:

> Contrary to what is sometimes said, this blessed disobedience solves many problems. In accumulating black follies and an experience which helps nobody, each generation blocks history so that we no longer see, in a moment, how to leave it; only children sometimes unblock the situation by seeing things in another way. Animals rarely disobey; genetic automata, some follow an instinct programmed since the origin of their species: that is why they have no history. We change, progress and regress, we invent the future because, deprogrammed, we disobey.[88]

This "deprogramming" is the relative neutrality of a ciphered zeroness that is articulable in code. It is the subject's casting off towards universality, in Serres's philosophy of the transcendental objective. It is the "bearing witness" we ascribe to information theory and quantum physics; that makes it a "literacy."

If Eve and Adam's frivolous act is considered as undecided in its moral nature, then those particular humanist positions would be mistaken: First, that humankind has been left alone in the world, with the sole and tragic spirituality of a regulative machinery (instead of an arcane architectonic body of laws) that operates obediently and reliably in official generality; and second, that this abandonment brings about the tragic condition in which the very possibility for comfort is a finite good that this machinery must administer to as best as it can. Because if Eve and Adam are absolved from the stigma of always already having submitted to corruption, then history does not serve to distance mankind from its lost original nature that was, purportedly, corrupt when history began. The sequential order now includes the possibility for regression just as much as that for progression.[89] Human nature now is not *good* nature; it is nature based on common sense, but "common" no longer carries an innocence. The spheres of "nature" and "value" are kept distinct now; they are kept apart by the encryptions and decipherments depicting the secret that is the universe's nature, those symbolic building blocks of the crypt embodying the law obeyed by the kind of universal human nature of which Serres speaks in the inchoative terms of *hominescence*. But if codes manifest those "building blocks" of the crypt, the *Grand Récit* that knows the age of the tree of life, what then are those "codes" made of?

Exodic Discourse

Beginnings and endings no longer amount to moments of immediate, full presence, of unobstructed lucidity and clarity, moments experienced only in the complete absence of doubt. Doubt comes from *duo habere*, as Serres notes repeatedly, literally meaning the cofactoring in of various things. Dating the objects of science in their incandescent, universal nature grants their existence a sense of being (in the double negation that constitutes the mathematical inverse) *not undecided*. This sense of the objective world is the intuition with which Serres's philosophy of the transcendental objective allows the subject of mathematical language to *rest* at noon when the light appears with greatest clarity; euphoric but also with historical doubt, capable of critique because equipped with the meteorological means of measuring the world through re-membering the age of the universe, the impersonal subject of mathematical language is after a new kind of knowledge: "My discourse is scientific and at odds with epistemology; it breaks with two millennia of method. Or rather, this old fiction is saturated with a different, incredible kind of knowledge. New knowledge. It is not fiction and not a true story I seek, but the exodic discourse."[90] Serres formulates an ambition with a maxim we have already alluded to several times; "if philosophy doesn't have to dominate science or become its slave or handmaiden, it must at least maintain compatibility with it."[91] His initial assumption is that this factual technical mastery has *expelled* us from the promise of a knowledge that could be acquired and shared equally, justly and uncorrupted, if only strictly rule-based methods are followed. Method, understood as "arriving at the best result with the least effort, as earning the maximum while paying the minimum," this constrains us to tautology, Serres maintains.[92] While "method minimizes constraints and cancels them out," he elaborates, "exodus throws itself into their disorder."[93]

The knowledge that derives from method alone, as arriving at the best result with the least effort, as earning the maximum while paying the minimum, maintains a *performative* and *immediate* relation to "the symbolics" that it applies in its formality: it assumes, ideally, the full presence of the elementary forms (or the words, or the thing itself, or phenomena) in the transparent neutrality of an ordered setup (axioms, the alphabet, a vocabulary, in numeration, or in the apparatus of optical physics). We could call this the epistemological assumption of existential quantifiers[94] and speak of *methodic reason's arithmetic of truth-value*: at home (in the ideal of such full and well-ordered presence) "right" and "legitimate" is the direction which is most economical in the sense of "earning the maximum while paying the minimum."

In contrast to such an *arithmetic* of existential quantifiers, the exodic knowledge Serres calls for looks at "the symbolics" with the eyes of the uninitiated stranger (as well as those of the estranged initiated) who finds herself, foreign and alienated, undecided, in the midst of places (mi-lieu) with none of which she fully identifies.[95] This stance also entails that knowledge must be formal, but it looks to information theory to provide an adequate philosophical stance.[96] Serres thereby foregrounds the most uncomfortable and, to many, outrightly *inacceptable* character of information theory: the physical theorem that information has its price. Subject to the thermodynamic principle of time's irreversibility, no observation is gratuitous.[97] The perfect experiment James Clerk Maxwell (and many others after him) dreamt of—an experiment whose agent would be a disembodied and pure mind, is impossible de facto (not by argument of principle), because it would have to assume an infinite amount of negentropy with which it could pay for the new information. Information can be acquired only at a cost—and this applies (to turn to what Brillouin and Leo Szilard demonstrated) *not* only to a subjective agency of understanding but also objectivity itself, via the ciphered code in whose symbolics "objectivity" is formulated. That information has a price is the theorem of a mathematical realism for Serres.[98] This is the reason that he is at odds with epistemology, which uses the balance between expense and acquisition in a non-stratified and global manner: best result with least effort, maximum effect with minimal cost.

For such a mathematical realism, nature counts as both rational (finite) and irrational (indefinite), as at once hidden (the encryption of the irrational) and exposed in the code as a sign of quantification (information as quantity), by the formal language of mathematics.

> Enlightenment philosophy teaches that the irrational must be driven out: what do the hideous statue and its inhuman form of worship have to do with us? But we have since learned to call anthropology what the Enlightenment cast out as madness or darkness, and we have also learned that exclusion brings us back to the sacred because the gesture of expulsion precisely characterises sacrifice. By rejecting this form of worship and scene as barbarous, we risk behaving the way the Ancients did. Therefore let's accept our anthropological past as such; ignoring it would make it return without our suspecting it.[99]

How then can we deal with the infinite (the mathematical irrational) from this stance of mathematical realism? What Serres proposes concerns the role of positions in a vicarious order: philosophy must learn to address the locus of the vicar, the joker, the general equivalent. Mathematics *is* the royal path to knowledge, he acknowledges, but the sovereignty of the mathematics cannot govern immediately. It would make irrational forms of worship return without our suspecting it. The sovereignty of mathematics can only be claimed from the position of expulsion from one's "proper" place, for the sake of a knowledge of universal nature.[100] He elaborates on the meaning of this for science:

> Nature is hidden [*dissimulée*] behind a cipher [*sous une grille*]. Mathematics is a code, and since it is not arbitrary, it is rather a cipher [*chiffre*]. Now, since this idea in fact constitutes the invention or the discovery, nature is hidden twice. First, by the cipher [*grille*]. Then with an ingenuity [*une adresse*], a modesty, a subtlety, that prevents our reading the cipher [*grille*] even from an open book. Nature hides beneath a hidden cipher [*grille*]. Experimentation and intervention consist in bringing it to light. They are, quite literally, *simulations of dissimulation*.[101]

Science, then, is the remembering of the locus of the vicar, the joker, the general equivalent. Its knowledge is not that of the architecture of the cosmos or of the world, but that of the architecture of the vicarious place. The agency of exodic knowledge is "exodic" because it is aware that it cannot "simply" (without cost, without sacrifice) identify with an assumed fullness of elements that were capable of rationalization (as epistemology arguably does, for Serres). It regards the symbolic nature of formal elements in mediate distance. It looks at elements as *cryptic* and *coded*, as encryptions of quantized strangeness, marked by chance-bound encounters (rather than as uniformly quantized in terms of belonging and membership). Serres wants to acknowledge "strangeness" as an *objective* quantifier in the domain of the vicarious, where it modalizes universal quantification: "Strange in its precise meaning, without relation to the judging subject, universal; strange, this means *improbable*."[102] To him, mathematics is a code that *keeps* the irrational (infinite) strangeness *contained* within the reality of rational actions. This allows us to think of mathematics as a means to maintain compatibility. One particular question clarifies the stakes of Serres's mathematical realism: "In other words, do our clear knowledge and effective technologies include dark patches of unexpected ignorance?"[103]

According to Serres's methodical reason in exodus (from a position of innocence), the symbolic nature of elements is held to be constituted by ciphers. From the detached viewpoint of the expelled mathematician, any finite group of ordered elements, whose combinatorial totality spells out the algebraic symmetry-space of an equation, is a cipher. What does an algebraic symmetry space mean? We can imagine an encrypted sum of transformations, in which for any given transformation that could be carried out there is also one that could cancel it (undo it). "Cipher" comes from *cifra*, for "zero" or "nothing." So understood, every code can be looked at as a cipher: the decimal number system using ten as logarithmic base for all numbers, just as validly as the hexadecimal number system, or the "myriads of myriads" of Archimedes' The Sand Reckoner.[104] To say that code matters is to say that every cipher introduces an encrypted rationality, a manner of counting that is, in each case, universally valid—but, because of the price of information, *not in equivalent manner*. We can no longer maintain, having generalized the relation of equivalence to one of equipollence (equality in power, force, or validity), that there is one particular scale which must count as elementary and principal for all things equally.

Why does Serres say that nature is "hidden twice"? Serres's first book was an extensive study on Leibniz,[105] for whom he maintains that, "the system of the models, this is a model of the system."[106] In his reading, Serres suggests thinking of Leibniz's universal characteristics not in terms of an ultimate ground and referent for systematic deduction, but in terms of mathematical translations, on the vicarious level of structures that build bridges between systems. "One needs to understand," Serres writes, "that in Leibniz's philosophy, systematic organization is ill-understood when thought of in reference to one sole principle of identity, in terms of univocity. Rather, it ought to be thought of in relation to all the variations which this principle accommodates."[107] This identity principle, he continues, affords rigorous treatment of "a content that is un-organised [*inorganisé*] by notions, and by thousands of possible architectures."[108] But this is not sufficient, he continues, because "it is impossible … to think one sole notion in an isolated and fragmentary manner, in its regional signification. It is essential for any one of them to be thought, defined, situated in reference to all the others. It is their nature to lend themselves for being integrated with their own *consequential* presentations— and in all of them at once."[109] There is, in effect, "always a plurality of possible deductive paths,"[110] and, "the multiplication of paths augments the richness of the analysis and the solidity of the connection."[111] Reason is addressed here, according to Serres, in terms of *richnesses* and *solidities* of different structures; this is a non-representational way to address Leibniz's distinction between necessary and sufficient reason. It is one that gives an "account" (*récit*). It addresses reason as a net of concurrent and multiply interlinked chains of deduction, in which values are polyvalent—"a web of many concurrent chains"[112] within which "one has to try to construct a network" that could "constitute the plan of the labyrinth."[113] Every approximation of such a plan "carries in itself a secret for which we, in our confusion, search."[114]

What is so counterintuitive from the epistemological point of view, which always assumes an ultimate referent code, is that this "secret" cannot in principle ever be disclosed and exposed, that the access to knowledge does not depend upon breaking an ultimate code by learning how it works in order to see what it conceals. It has long been clear that the endpoint of a methodical and deductive path, followed rigorously, starting out from elements deemed principle and ultimate, always proves to be the product of assumptions written in from the beginning. To pick up a famous example: if a computer simulation is to calculate the limits to growth, it will always produce a particular value for that variable around which the system is set up: it will never say that the assumptions that went into the formulation of the simulation are wrong.[115] Whereas for methodic knowledge "at home" there is only tautology, exodic discretizations and distributions, the adoption of strangeness as an objective quantifier that modalizes universal quantification allows the factoring-in of the unknown in a way that need not reduce it, whether positively or negatively. By acting on ciphered equations, such structures of deduction (for which the model of the system is the system of the model[116]) conserve the unknown as a secret, as an element of chance and originality in the cipher code they apply.

Serres's interest in cryptography is one that gives primacy to the processes of translation between what is technically called "source text" and "cipher text," and it informs the role of "importation" therein (as we will discuss in Chapter 7). Translation is stripped of dependence, however light, on an original source text, and it is shewn of the illusory certainty that any one particular cipher text must provide the ultimate key to access the source text in its full extent. For Serres's exodic knowledge, mathematics is the cornucopia from which wells the richness (or scarcity) in terms of which an analysis can be meaningful, or a connection can be solid.

7

Sophistication and Anamnesis: Remembering an Abundant Past

Architectonics within the Domain of a Withholding Power

The Price of Truth, and the Price of Information

"We may ask if the whole of the enormous movement of the modern economy—what is now a global production machine—might not be the last and most radical way to eliminate the gods, to do away with gift-giving and debt," writes Hénaff in his 2003 book *The Price of Truth*.[1] He continues, "It may be that we produce, exchange and consume in order to reduce our relationship to the world and to each other to the management of visible and quantifiable goods, to prevent anything from escaping the calculus of prices and control by the marketplace, so that the very concept of the priceless would finally disappear."[2] This outlook provides the right frame for opening this chapter. There has been one key notion that especially haunted the research behind this book, and this is the concept of the *Price of Information*. It appears in many places throughout Serres's books, but in greatest detail and most explicitly in *Hermes*, volumes II and III.[3] It is introduced there in a technical sense that (as we have seen in Chapter 2) entails a particular economy of thought which results from applying the communicational physics point of view on information theory, as Serres develops it. Why does this theorem speak of a "price," and how is that significant for our context at all? A price is a number that determines the worth of something within a continuum of differences that are infinitesimally "resolved," made "compatible," "proceduralized" in a variable (temporary and circumstantial) manner. The relevant question for our discussion now is: Does the theorem extend such "economical thought" beyond all bounds? Must life itself, with microscale biology, with the genetic code, be recognized to 'negotiate' the potential-consuming and -producing orders of information?[4] And does this not mean that the concept of the priceless has finally disappeared? If this were the case (so Hénaff elaborates with respect to his own approach to an economy of truth), then nothing would remain outside the realm of commerce. Material innocence would finally have been achieved: no more faults, sin, gift-giving, or forgiveness, nothing other than mistakes in calculations, positive or negative balance sheets, and payments within agreed deadlines. This seems to be the world that is in the process of emerging in our ordinary practices of production and exchange.[5]

This chapter will discuss how we can think of the relation between knowledge and currency through the idea of sophistication, and the quantification thereof. Knowledge with regard to *sophistication* (in contradistinction to *sophistry*) thwarts the categorical distinction between things that have a price and things that do not, without abolishing the distinction itself; rather, knowledge articulates architectonically a space of *withholding* where the concept of the priceless neither dominates nor has disappeared. To maintain this position, the chapter will discuss sophistication with regard to *anamneses*, as a remembering of an abundant past that promises a novel— an immanent kind of—understanding of classicism. It will establish that the link between commerce and communication introduced by the theorem of the Price of Information does not raise a new problem, but rather one at the very heart of how commerce has been dealt with in philosophy all along. It is helpful to recapitulate this history.

Hénaff's book is, overall, set up as a reconsideration between the two competing legacies, that of the sophists (which we could call that of humanism, taking man as the measure of all things) and that of philosophy in the tradition of Socrates, Plato, and Aristotle (which we could call that of *being*, ontology). What is at stake between these two legacies, Hénaff maintains, is the question of the priceless; the question, he adds, that we can no longer evade. But how could Hénaff possibly approach the question of the priceless from a discourse that is not ultimately itself derived from, and hence incapable of transcending, either the humanist discourse of the sophists, or the metaphysical one of the philosophers? Hénaff responds to this problem early in his introduction: "How are we to understand the phrase 'the price of truth'? It ought to be enough to say that we should understand it literally. This means that we are not employing the usual metaphoric sense of the phrase, which refers to the moral cost of the effort or renunciation required to proclaim, discover, or confess truth."[6] What I want to demonstrate in this chapter, with reference to Hénaff's *price of truth* in relation to Serres's *price of information,* is that a physics of communication (indeed as Serres also calls it, the "*mathematisation of empiricism*")[7] opens the way for a science of "literalness" that we find at stake in Hénaff—a science that deals *quantitatively* with sophistication.

Let us consider what is at stake with the price of truth taken in its literal sense, by Hénaff. He writes,

Whether the question of truth concerns self-discovery … or finally, an investigation to establish the facts … it always involves an aspect of reality that is either misunderstood, difficult to grasp, or obscured by some form of resistance or refusal that needs to be overcome. In this case the "price" of truth designates the honesty or courage required to proclaim or recognise proven facts; in other words, to refuse to lie, whether it be in research, a confession, or an investigation. It is a spiritual price, a symbolic price.[8]

A science of such "literalness" would be a science of a "spiritual" or "symbolic" economy. It would be a science of an economy that does not seek to "do away with gift-giving and debt," an economy that does not seek "to eliminate the gods," an economy that does not subscribe to the self-righteous hubris which holds the promise of order based entirely upon "material innocence."[9] We can easily recognize that the very concept of Serres's book *The Parasite* (1982) can be read as a response to any of those aspects as well. It is the book where Serres proposes for the invariant motive of the priceless this "eventuality" within a "generalized equatoriality," the idea of the incandescent Paraclete that we have elaborated in the last chapter. For now, the predominant question will be: How could we possibly speak adequately about a spiritual or symbolic economy *in terms of a science*? Let us begin with a closer look at the problems a writer faces in relation to the question of the price of truth, before coming back more strictly to the philosophy of Serres.

A Science of Literalness, and the Convertibility of Truth

Hénaff's quest was, quite simply, "to investigate into the specific ways in which philosophy has dealt with the question of money and venality."[10] Plato witnesses a radical transformation of the concept of truth, *alétheia*, a term that has a long history in ancient wisdom and legal practices, and which is usually translated as the unveiling of the hidden, the movement out (*a-*) of the hidden (*létheia*). When the question of money assumes a key role, this concept begins to be transformed. I will briefly outline the shifts of this turning point based on Hénaff's own book, as alongside Marcel Detienne's classic work, *Masters of Truth in Archaic Greece* ([1967] 1996), on which Hénaff also draws. Our focus of interest thereby will be to trace Hénaff's search for hints that allow him to substantiate his abstract notion of "literalness" in terms of a symbolic or spiritual "price" by studying the role of money in relation to this major turning point that is (more often than not) studied with an interest in the role of the Greek vocal alphabet alone.

Before truth became an essential philosophical concept with the pre-Socratics, it often appeared allegorically in the oldest texts on wisdom and its practices by seers, magi and, above all, poets. It was consistently associated with *mnémosyné*, memory, which has less to do with the ability to remember than with a visionary power that provides access to the past, present, and future. In Greek thought, it was pictured as the "Plain of Aletheia," as the visionary locus where the gods and heroes resided: poetic language, with its visionary and inspired verses, was thought to provide access to the divine realm. "Poetic language has its effect in producing what it states. Without it, the hero cannot attain the glory granted in and by the song that brings heroic actions into the light of *aletheia*. The hero becomes what the poem says about him."[11] The performativity of language not only concerns songs of glorification but is also evident in other domains, primarily in divination, where oracular speech accomplishes what it announces, and in law, where whoever has the authority to say what is just—for instance a royal figure—makes it so by pronouncing the diktat in speech. Speech in this framework was a performative kind of speech that belonged to the forces of nature, speech as a physical reality. As such, it has poetic, divinatory, and legal powers. With the emergence of the city, this configuration changes. The polis takes shape with the institution of equality among warriors (Hoplite reform). A neutral space is designated in the center of their circle for the common spoils to be distributed by lots. This place is the origin of the public space. It becomes a space of debate, the center from which everyone must talk to everyone else. "Magical and effective speech that creates what it states is replaced by dialogue and contested speech. Ordeal gives way to investigation and the requirement of proof. Authoritarian discourse is replaced by the exchange of arguments. Speech loses its status as full substance, as effective power. Separated from its effects, it becomes autonomous form and instrument."[12] It is in the course of these developments that the notion of truth began to transform: opposed to *aletheia*, which is uttered by a sacred mouth, is the *doxa*, which is "the knowledge appropriate to a given situation and accessible to everyone."[13] A split between statement and the act of stating is thereby instituted, between the world and its representation, image, and reality. Simonides of Ceos, who is according to Detienne the first poet to illustrate this change (a process that Hénaff identifies as one of "secularization"), not only breaks with ritual incantatory poetry and the performative conception of language, he is also said to be the first to compose poems for a fee. The old *sophos*—the seer and inspired poet—turns into a language professional. The mastery of the art of speech is increasingly understood as a craft comparable to that of many others, including that of physicians, musicians, weavers, or architects. All of them received wages in exchange for their services. This is how the new speech professionals brought *aletheia* to the marketplace and began to set a price on it. With this, *aletheia* ceases to be the sole property of the *sophoi* and poets. Together with the emergence of mathematics, credited to the physicist-philosophers like Pythagoras and Thales, a different conception of truth comes to prevail: the objectivity of inherent relations between things themselves. This objectivity can be reached through cognitive discipline. Knowledge can be passed on through public education rather than through secret procedures of initiation, as Hénaff puts it.

What we witness here is the bifurcation of two different positions regarding the question of truth: the hermeneutic tradition of decipherment now confronts a new tradition, in which truth is viewed as objectivity, as the foundation of an agreement. "What is now at stake is less discovering truth than constituting it."[14] Mathematical or physical knowledge, the logical arguments in which such knowledge is formulated, is neither sacred nor initiatory. It can be acquired in the same way as any know-how: through techniques of learning. What used to represent a strictly spiritual power now became an instrument of mastery, a way of earning a living, a source of wealth. At the same time, the new type of philosopher continues to enjoy the social prestige of the old-fashioned wise man and takes advantage of it. The sophists introduced "the power of *pseudos*"—the feigned, the insincere, that which is only apparently what it claims to be—into philosophy. They use the logical forms as a mere means of arguing. They invent games of *logoi* in which the question of truth—the being of the thing—becomes arbitrary. Much like merchants, they no longer need to know the nature of the product.

Like a merchant, they are mainly interested in convincing their customers to buy it. The art of hermeneutic decipherment gave way, through the very means by which it came to be propagated, to the opening up of a realm of illusion and falsehood—and it is not the moral falsehood or malice of an individual subject that is at stake here, but, as Hénaff puts it, "ontological deceit."[15]

The interesting aspect that appears now, from Hénaff's discussion of the bifurcations and transformations of the concept of truth, is another bifurcation that goes along with the first one: "when we deal with the critique of money and the denunciation of the harm caused by venality or corruption ... we leave the realm of tragedy."[16] When something of the order of "ontological deceit" is at stake, we are no longer within the purview of some collective banishment of the *tragikos*, the scapegoat. This is no longer an adequate manner of dealing with the situation. Money is not epic. "Money as such"— that is, when it is not allied immediately with the classical forces of evil—"seems to belong to the reign of *lowly* things."[17] Money, viewed like this, figures vis-à-vis wealth in general, which seems not always to have been synonymous with money, and which moreover could have a noble and enviable status. Now this is indeed interesting: if money, and the powers of the *pseudos* that connected it with the concept of truth, constitutes an alternative realm to that of the tragic, then we may perhaps find here a different approach to history than those which understand it as accounting for (myth), and bearing witness to (Christianity) the present of sacred violence in the expulsion or sacrifice of the victim for the sake of the peace of the community. Hénaff's quest to trace a literal understanding of the price of truth follows figurations of *contempt*, of despising what is powerful without claiming dominance and economic and political mastership. What we will be interested in is how this "despising," which is linked to Hénaff's notion of ontological deceit, can be neutralized into the mere "keeping of a distance" facilitated by a science of such literalness. The keeping of a distance ought to be possible through considering the power of the *pseudos* as a withholding power, rather than as a deceiving one. We will argue that architectonic articulations give this power a public face, and thereby establish the distance between deceit and withholding that characterizes the domain of the pseudos. We will be interested in setting up a notion of metaphysical import in the place where there is ontological deceit in Hénaff.

In Serres's white metaphysics, categories are absolute, (not exhaustively reasonable) and yet they are so in relative fashion—relative to the group concepts in which they feature, and from which they are being sourced (Chapter 5); categories are "modular," they are absolute and yet relative to a scalarity, a modeled "one-over-many" relation that needs in no way be "typical" (relative to a particular order). In short, categories are architectonic in the sense that they require mediation by structures—like this, the classification they facilitate can count at once as universal and yet capable of addressing singularity. How could we possibly speak, adequately, about a spiritual or symbolic economy in terms of a science, we asked at the beginning of this chapter. We saw how the sophists invented games of *logoi* in which the question of truth—the quality of being of the thing—becomes arbitrary. They do so much like merchants, who are sophisticated to a degree, where they no longer have any interest in the nature of the product. Undoubtedly we are living today in our dealings with data with a resurgence of such sophistry and sophistication. Our question now is: How can such an economical understanding be anything else than "ontological deceit"? We want to see more closely now how the concept of "structure" plays a crucial role therein, and how with it, the games of logoi can be addressed in their architectonic make up. We will see how there is an understanding of architectonic classicism with which it becomes possible to remember an abundant past as a means to keep futures open.

Classicism: Remembering Contemporaneity

Classical Analysis, Symbolic Analysis

Let us begin by picking up Serres's view that there are dark meanings that are not only the symbols of history, but that also pertain to a reasonable, scientific order.[18] How can we think of this breath, of such a soul with which one can get in touch, in a reasonable, scientific order that is articulates as the domain of a withholding power? We must think about structure as the facilitator of metaphysical import: the spiritual, the symbolic economy, in Serres, is one of *metaphysical import* rather than of *ontological deceit*: structures are keys, I will maintain, but they are not "the mysterious key that would open all doors."[19] Rather, we need to think of structure as "a methodic, clear, well defined and elucidating concept,"[20] a concept that grasps confusing situations with clarity. A concept aligned with rational sobriety, but which is also poetic, dramatic, in that it is endowed "with a normative, a cathartic and a purgative definition."[21] Surely, if structure were merely declared now to constitute the novel stage for tragedy and comedy within an epic scope (a scope of heroism), it could not count as "a methodic, clear, well defined and elucidating concept." We will see how in Serres's take on mathematics as a language, mathematics that is universal and objective yet without erasing positions that need to be occupied by a subject, the former role of heroism is substituted by a different organization of potency. Serres's concepts, in the terms of white metaphysics, are polynomial concepts, concepts that incorporate multiplicitous names, concepts whose polynominality is derived from the world's proper name, the sixfold Panonyme: *Pantope* (all the places), *Panchrone* (all the durations), *Panurge* (universal worker), *Panglosse* (all the languages), *Pangnose* (all knowledge), *Panthrope* (all the sexes). Let us try to work out what this entails.

The question that Serres inherits from his teacher, Bachelard, is the quest for a notion of form (rationality, classicism) that does *not* have to be bought at the price of relinquishing the expression of a cultural content in its obscurity and endowment with arbitrary customs, in its confused messiness; in short, in its *reality*. Bachelard worked throughout all his books on what he called a new scientific spirit [*Nouvel Esprit Scientifique*].[22] Serres is in close agreement with this idea when thinking about a history of ideas that is, at the same time, also an idea of history. Bachelard sought to develop an epistemology of his new scientific spirit as a kind of a psychoanalysis of the objective world. Like him, Serres is concerned with a certain kind of essentialism, but for Serres it is a mystical rather than a mythical essentialism (a *substantial* notion of identity is central in Serres white metaphysics, it is why becoming universal is becoming no one in particular, it follows the path of dedifferentiation). He seeks not to represent such essentiality positively for *reflection*; rather, the transcendental objectivity he pursues seeks to model it for the sake of *communicability, mobility* and *instrumentation* (cf. Chapters 3 and 4). We can think of such modeling as the imaginative conception (hence rather, "inception") of ciphered keys in the style of a Rosetta Stone, that encrypt into decipherable geometrical statements some of the abundant reality of an experience-relative universal order. The instrumentality of such "inception" is what I have described as that of a cornucopia. Serres's is a way of reasoning that in its methods neither positivizes nor negativizes the grounds of its own reasoning. These grounds, to Serres, are always those of experience, and experience can be the subject matter of science only insofar as it is not "personal," as it could "happen to anyone." We have seen in Chapter 5 how Serres considers his philosophy as distanced from methodically grounded reason. The difficult gesture to grasp with regard to this distance is that it entails a kind of reasoning that conspires with *episteme*, while being at odds (also in the sense of being "awkward," cf. *Le Gaucher Boiteux* (2015)) with epistemology. As for Bachelard with his notion of an objective unconscious, there is an unfathomable depth to experience for Serres. But this depth comes from the role that death plays in experience; death as the objective and ageing nature of the universe. This introduces an anonymous, impersonal and objective dimension as well, just as does the unconscious in Bachelard. But for Serres, this depth is proper to the real universe itself (not only relative to psychological, subjective faculties). Forms are the bearers of rational regularity, and they originate in death, he maintains. This is the idea which orientates his proposed mathematical mysticism, which is nevertheless as "realism." It is crucial for understanding why an objective kind of convertibility (as Hénaff seeks to foreground within philosophy with his question of the price of truth) is transposed by Serres into the registers of physics itself: as the Price of Information (cf. Chapter 2, "Quantum Literacy").

With Bachelard, Serres shares the ambition for a new *classical* philosophy, one where not only sense and signification count as real, in their culturally biased ways, but also one where truth would again be important. Where, for the merchants among the sophists, a notion of pricelessness would again be important, in a symbolic economy where pricelessness would manifest in knowing more about the "goodness" of a tradeable good (in the double sense of "trade" and "tradition"); this goodness becomes quantifiable and is what his architectonic approach to mathematics and information articulates as the domain of the power of the pseudos, which turns thereby into an indeterminate power of withholding. Within Serres's communicational physics, the worth of a thing is kept in oscillation between its articulation as a product and the communicational *nature* of it as a product (rather than as worth relative to an "ontology"). The nature of a product is to be investigated empirically and mathematically. Nothing less is at stake than the idea to make culture the object of science. Serres sets out in the article entitled "Structure and Import: From Mathematics to Myth" (1961), how "Romanticism ... introduced the project we are still in today," namely, "to grasp the plurality of significations and to decode all the languages which are not necessarily those of pure reason."[23] It is with what he calls the technique of symbolic analysis that Romanticism has invented a method to pursue this path. Serres offers a concise formulation for this claim:

When the problem of classicism is truth, and the domain of application of this problem is reason, then the problem of the Romantics was the question of sense, and its domain of application was the historical totality of human behavior [*attitudes humaines*]; the methodological horizon of the former is that of order (of deductions, themes, conditions, etc.), the methodological horizon of the second is that of the symbol.[24]

It is with a view toward this "historical totality" that we must relate sophistication to *anamnesis* in order better to grasp Serres's maxim that models are not primarily to be representative or explanatory but *instrumental* and *operative* (they "realize" a phenomenon) and also *active* (since he attributes mentality and character to instruments that operate, as we have seen). Models are required to facilitate a taking and a giving for both equatorial poles of Serres's operational principality of reason (which is "that there be no reason," see Chapter 3), the equipollence between the real and the rational. Models subject to such a principle of reason convey "dark meanings" in "transparent structures."[25] In this sense they provide for a scenography: they facilitate an order of profiles ("des attitudes humaines") in relation to other profiles ("des attitudes humaines"). This order is to be called an architectonic order because the profiling of such scenography (in architectonic registers: scenographia) happens not only immanently (among scenographic profiles) but also in relation to the reality of a common ground (in architectonic registers: ichnography) as well as to an explicit (transparent, ciphered) mode of notation (in architectonic registers: orthography, as in columns and their orders). With this architectonic approach, to which we will come back more extensively in the "Coda" to this book, Serres introduces convertibility as the neutral element (Hénaff's price of truth that is to be assessed relative to the Price of Information) into the modeling space of geometric reasoning. This is how and why Serres notion of models is a mechanistic notion, one which thinks of mechanics as an art. His concept of the model installs an inverse operation to the Romantic approach, which had treated the inexhaustible totality of worldviews only *symbolically*, by rooting them in archetypes, but not *geometrically*. The inverse operation, with which he conceives of models as mechanical instruments, not only deploys a geometrical method for the symbolic treatment of the inexhaustible totality of worldviews but also introduces the economic possibility of import and export of "truth" (as encrypted meanings) between symbols and the archetypical grounds in which they are to be rooted (according to the Romantic treatment). Accommodating this neutral element of inverse operations which makes the models mechanical and instrumental, Serres's concept of modeling adds to the symbolic analysis by which the Romantic models proceed; it adds that models now can make "dark contents" graspable through "transparent" (because ciphered) structures. Such structures *model* the procedures of symbolic analysis. They are yielded via an operation that works *inversively* to the ones operating with archetypes; the latter had, arguably, provided for the Romantics their immediate principle of reason. For Serres, the symbolic provides for an economy of metaphysical import, not primarily an analysis. Therein, archetypes provide legendary grounds of traces (ichnographies), and form one particular source among others for a science of history. The history of mathematics is another particularly important source of such "wealth" in Serres symbolic economy: "mathematics, with its expression through formulae, is a theory which is open towards the inside and closed off towards the outside."[26] While it may appear to anybody but mathematicians as a "normed universal," nonetheless to mathematicians themselves, the formulaic system which they inhabit is actually "an open field."[27] For the mathematician, Serres elaborates, "there is not only the necessity to complete his building, but most of this exhilarating laxity of the not-completed, of that which is only insufficiently closed, and the urge to take up work on that same building once more, and then again once more."[28] In Serres's symbolic economy, it is not only the archetypes of myth that feature (as ichnography) but also mathematics itself (as orthography): it is, indeed, how he can speak of *anamneses* (in the plural), where Bachelard reverts to the subjectivist legacy of an unconscious. Serres performs a clear transposition here from epistemology to architectonics: what Hénaff calls "games of *logoi*" as the Plains of an open Field is being transposed to Serres's mathematical "*anamneses*," that invite the "building" of such plains, of such fields. As a result, what appears to be ontological deceit from an epistemological point of view can be considered in terms of metaphysical import and export from an architectonic point of view. Serres elaborates with regard to mathematics:

> Together with the universal possibility of translating themes, there is also the universal possibility of letting them vary within themselves, such that their genealogy can be recognized. Formal thinking readopts history and endows it with a spectrum of meanings, that is, the totality of imaginable meanings. From the point of view of the system, all meanings belong to history; from the point of view of history, the system at once makes sense and is infinite.[29]

In his understanding of models (that are not to represent but to realize), Serres couples the "totality of all worldviews" (a mythical method that proceeds by means of archetypes for the Romantics, ichnography) with a "totality of all imaginable meanings" (Serres's mathematical *anamneses*, orthography) in the same manner as he couples the real and the rational. Both totalities are set relative to each other in a *projective* equatorial relation of equipollence that needs to be reasoned by means of analysis relative to the architectonic profiling (scenographia) of a statement in terms of the modular order as equality in validity, force and power. If the formula of architectonic formulation is reasoned, it produces a demonstration as well, as we would expect within the horizon of epistemology (and its "methodic" reason). But with Serres's exodic reason, it remains open what these demonstrations "demonstrate." Exodic reason can never be sure of itself. It produces *architectonic* demonstrations, not ontological ones. They demonstrate the truth of this equation in its relativity to the "instrumented" terms of one particular act of sophistication. Such architectonic models enable through structural methods a comparative kind of empirical investigation of the archetypicality that "grounds" symbolic analysis. Let us try to trace how this can be thought of, by beginning to state Serres's guiding assumptions explicitly: the methodical horizon of classics is provided by a notion of order, he maintains, while that of the Romantics is provided by symbols. Order, in most basic form perhaps, arranges elements in ranks, lines, and classes. Symbols, in the most basic form, present what counts as an element from the point of view of order as broken, as artificially, contractually, fitted together. Elements that are considered as symbolic provide an overabundant stock of combinatorial, possible "fittings" (permutations): it is why a limited stock of characters in an alphabet can yield an uncountable number of words (that might at least potentially make sense). The relation between order and symbol is what animates, what vivifies language. Moreover, with Serres's communicational physics, we can think of this relation between order (classicism) and symbolics (Romanticism) as *metabolic*; as co-constituting the metabolism of a language, as an organizational constitution that settles, in entropic/negentropic terms, the relevant factors that keep a language alive (or not) (cf. Chapter 2, "La langue est une puissance"). Furthermore, Serres maintains, if one wants to be faithful to an ideal of order it is sufficient that there exists an ideal model that realizes the ideal order completely and exhaustively (*parfaitement*).[30] Such a model is provided, classically, by the exact sciences. It is what the diverse guises of classics liked to call a mathematical order. From the point of view of the Romantics, this order too was archetypal, because mathematics hinges equally on symbols. But Serres's approach does not seek to be faithful to an ideal of order; it is committed to realism. Symbols, in mathematics, serve to provide for classical science what Serres terms "a distinguished model" (*un modèle choisi*): a distinguished model is a model elected with respect to order as an ideal, but in consciously maintained distance to it.[31] As soon as one extends the domain from pure theoretical mathematics to the novel fields of its applications, a mathematical notion of order in a pure sense presents itself as a hindrance: while mathematics literally exploded into novel levels of abstraction in the nineteenth century (and ever since), what we call today the science and the philosophy of the Romantics recognized the ideal of a pure mathematical order as inadequate not only for "the dark contents" to be examined in culture, but (as Bachelard would stress), also for the novel domains of scientific analysis itself (electromagnetism, chemistry according to the periodic table, statistics with regard to mechanics, economy, society, electricity, but also novel infrastructures for signal transmission, radiophone, photophone, photography, etc.). The true scientist, Serres believes, will readily accept a certain kind of darkness of meaning in those fields: "Rather than relying upon an ideal model which serves as a normative index, one now has to construct a concrete model within the very domain being analyzed, and then to refer to its content rather than its order. The content then no longer imitates an ideal model, but repeats, content for content, one concrete and universal symbol."[32] There is a passage from order to content, Serres and Bachelard agree. But according to Serres, Bachelard remains stuck within what he calls the method of symbolic analysis, whose heirs are the Romantics. Let us try to clarify the lines and legacies.

What ultimately distinguishes the two approaches that are being profiled here, the classical and the Romantic, is their respective understanding of the abstract and the concrete. While the classical mind-set maintains that the only thing that is intellectually graspable is the ideal of an order, and that it is this ideal order alone that counts as generous in the sense of bringing steadiness, stability, and peace, it is this classic notion of order (as abstract, mathematical) that the Romantic mind-set stigmatizes as hegemonic. Serres applies criticism to the romantic's own conception, according to which they lay claim to more palpable (if somewhat short of complete) access to the symbolic terrain than mathematics can envisage. He writes that with their accusation of abstraction with such hegemony, "symbols have descended from heaven to earth—even if not entirely so, because they have descended only from the heaven of ideas to the earth or the history of myths."[33] What is happening here, Serres tells us, is that the pure (classicism) transforms itself into myth.[34] Myth, for the Romantics, is now at once universal and singular:

"It is in this sense that the technique of analysis of Hegel, Nietzsche, and Freud is symbolic and archetypical: The only question was to know where to find the archetype, from which symbolic totality one is to take it."

And, furthermore,

> many of the symbolical analyses of the nineteenth century chose their models in the history of myth for this: Apollo, Dionysus, Ariadne, Zarathustra, Elektra, Oedipus etc. They all represent eminently (that is to say, they symbolize) the essential totality of a culturally significant content. The meaning of this content is being grasped whenever one can demonstrate that it picks up the archetype, that it repeats and re-actualizes this archetype, that it guides it from myth to history, from the eternal to the evolutionary.[35]

A sort of productive self-reference is introduced in this period, as a strange kind of property of symbolic elements (elements that never come without being symbolized), dependent on thinking in terms of archetypes and a notion of conception reasoned through representing it (the ideal order of truth in classicism), conjoined with a notion of infinite, energetic, *working through* of such reasoning (method of the Romantics). Serres elaborates, "There is a correspondence of equal sense between the content and its symbol. And this correspondence engenders history, or the eternal return."[36] The essential thing for Serres about this profiling is that with this passage from classicism to the Romantics, "the notion of the model goes from clarity to darkness. With regard to problems this means: from the true to the significant. From the normative to the symbolic. From the transcendent to the original."[37] However—and this is where Serres's approach to archetypes via architectonics becomes salient—myth is not only a symbolic origin but also the *original* origin. It is the first beginning. As he sees it: "This short analysis shows us the concepts according to which we have been living until yesterday: the problem of sense and sign, symbolisms and language, archetypes and history, the grasping of dark cultural contents, fascination with the original and origination, and so on."[38] The architectonic reasoning he favors sets in after this: "But we need to emphasize the variation of the chosen models. What we did not know then has meanwhile dawned upon us as bright and clearly as a thousand suns: when we vary our problems, then we also vary our ground of reference."[39]

The novel classicism at stake for Serres is one where the grounds of reference, where the reason that supports reference, is taken into account *critically*. And this means that the notion of structure cannot count absolutely as "the mysterious key that would open all doors."[40] The symbolic analysis of the Romantics, hence, "is not at all a methodical miracle of genuine character. It is merely one stage within a variation."[41] Before continuing this adventurous tracing, let us look at how structure is seen by Barthes, in his discussion of the Eiffel Tower as a modern myth; as indeed just such a romantically fallacious mysterious key. This will help us to profile more neatly Serres's own idea of structure which remains miraculous, remains a key, but reasonable like the Rosetta Stone. That is, cryptographical, not mysterious.

Interlude: The Eiffel Tower, Archetypal Symbol of Existentialism?

Barthes begins his text,

> "*The Tower is there*; incorporated into daily life until you can no longer grant it any specific attribute, determined merely to persist, like a rock or the river, it is as literal as a phenomenon of Nature whose meaning can be questioned to infinity but whose existence is incontestable."[42]

He continues,

> It will be there, connecting me above Paris to each of my friends that I know are seeing it: with it we all comprise a shifting figure of which it is the steady centre: the Tower is friendly … a universal symbol of Paris, it is everywhere on the globe where Paris is to be stated as an image.[43]

Furthermore,

> its simple, primary shape confers upon it the vocation of an infinite cipher: in turn and according to the appeals of our imagination, the symbol of Paris, of modernity, of communication, of science or of the nineteenth century, rocket, stem, derrick, phallus, lightning rod or insect, confronting the great itineraries of our dreams, it is the inevitable sign; just as there is no Parisian glance which is not compelled to encounter it, there is no fantasy which fails, sooner or later, to acknowledge its form and to be nourished by it.[44]

In this characterization, the tower is modern for Barthes in the sense that it has a function, but the function it has is itself subordinate to, derived from, a myth that it continuously recreates: "the Tower which will appear, reduced to that simple line whose sole mythic function is to join, as the poet says, *base and summit*, or again, earth and heaven."[45] In this sense we can think, with Barthes, about the Eiffel Tower as a myth—but one that is *modern* (not traditional, or "romantically neoclassical"). It is a modern myth in the sense that it performatively enacts the originality it transports through time. It is a young myth that cannot age. Such a modern myth is truly a myth of the day, a myth of the everyday.

Building a Cipher

As a symbolic function, the tower for Barthes is a cipher. Its functioning is mythic in the sense that it "incorporates an infinite circuit of functions."[46] He elaborates, "This pure—virtually empty—sign—is ineluctable, *because it means everything*."[47] There is a clear moralization in the account Barthes gives of the tower. He calls it "friendly"[48] and a symbol that "offends nothing in us."[49] This is the crucial point at which we can begin to see where the transition lies from Barthes's symbolic cipher (in terms of what I want to call here *structuralist existentialism*) to Serres's objective cipher (the objectivity in which nature is "hidden twice," by a cipher (*la grille*) and by a "hidden cipher" that concerns sophistication, as in dexterity, modesty, and subtlety (*une adresse, une pudeur, une subtilité*)) that Serres wants to propose for his realist classicism. Different from Barthes's symbolic cipher, Serres's objective cipher entails a notion of the object in terms of its quantified strangeness (see Chapter 2), whereas for Barthes, it is all about exposing the tower as an object that is *comfortable.*[50] This indeed appears like the straightforward contrast to the notion of the object Serres develops, for which all depends upon the object's status as being *not-entirely-ordinary.* What is not ordinary is challenging; it is uncomfortable; it is not "friendly" but "strange," potentially offensive, demanding, but also potentially rewarding, granting, giving; and, as we will see shortly, the strange object is not only abstractly generous or reductive (abstractly giving through delimiting, as idealist classicism maintains) but also receptively generous, concretely giving (through being generous *in how* it is being receptive, as Serres's realist classicism maintains). The object notions in Barthes and Serres are indeed opposed in how "comfort" is to mean smooth functioning, with or without resistance, with the aim of gratuitous accommodation or not, and so on. That is, they are opposed in terms of relative valuation, rather than simply contradicting each other, because there is little doubt that Barthes conceives the object on the same level of unsettled mobility as Serres does. But this unsettled mobility for Barthes's structuralist signification is not relative to architectonic models (in the plural, for Serres); his architectonic model (singular, mythical) is an absolute cipher, not an architectonically reasonable one. The tower belongs to the universal order of tourism, Barthes says; "[it] transform[s] the touristic rite into an adventure of sight and of the intelligence";[51] it is a mythical object because it "first is of a technical order,"[52] but as such, the tower has "provisions," no utility or use. For Barthes it is the monument of "the gratuitous meaning of the work."[53] The tower "offers for consumption a certain number of performances,"[54] but (and this is why it counts to Barthes as a modern myth) it also provides for the "demystification of how it is made"; it demonstrates a "spectacle of all details."[55] The Eiffel Tower affords its visitor "a whole polyphony of pleasures, from technological wonder to haute cuisine, including the panorama,"[56] and for this reason, "the Tower ultimately reunites with the essential function of all major human sites: autarchy."[57] Self-founding, without reference to anything else, "the Tower can live on itself: one can dream there, eat there, observe there, understand there, marvel there, shop there; as on an ocean liner (another mythic object that sets children dreaming), one can feel oneself cut off from the world and yet the owner of a world."[58]

This latter part is somewhat discordant now: The tower would only be a site of consumption, and "comfortable" in this very sense, if no inclination toward mastery was at stake. And Barthes is clear about this: "one can feel oneself cut off from the world and yet the owner of a world."[59] He regards it as a modern myth, and a modern man, to him, wants to be autarchic: he can appreciate the tower because, "man always seems disposed—if no constraints appear to stand in his way—to seek out a kind of counterpoint in his pleasures: this is what is called comfort." In short: "The Eiffel Tower is a comfortable object."[60] There is a dialectical learning involved in the scenario Barthes sketches. There is the idea that structuralism offers the chance for learning on an egalitarian basis This is undoubtedly the great beauty that graces Barthes's text. An architectural object—to the degree that it has no use—is actually able to "transform the touristic rite into an adventure of sight and of the intelligence."[61] How? "[T]he panoramic vision added an incomparable power of *intellection:* the bird's-eye view, which each visitor to the Tower can assume in an instant for his own, gives us the world to *read* and not only to perceive," Barthes maintains. It "corresponds to a new sensibility of vision"[62]—he is also looking for a novel sense of intuition. For Barthes, it is a sense of intuition that, "permits us to transcend sensation and to see things in their structure … intelligible objects, yet without—and this is what is new—losing anything of their materiality."[63] With this, he asserts, "a new category appears, that of concrete abstraction."[64] He too is looking for a neoclassical philosophy, one that could reconcile the abstract with the concrete, a way of transcending the contrasting approaches of classicism and Romanticism detailed above. What he finds manifest in the Eiffel Tower is a real referent for the problem of the "reference base" that we identified; the awareness that whenever we "vary our problems, then we also vary our ground of reference."[65] Barthes elaborates, "every visitor to the Tower makes structuralism without knowing it,"[66] and furthermore:

> Paris spread out beneath him, he spontaneously distinguishes separate—because known—points, and yet does not stop linking them, perceiving them within a great functional space; in short, he separates and groups; Paris offers itself to him as an object virtually *prepared,* exposed to the intelligence, but which he must himself construct by a final activity of the mind: nothing less passive than the *overall view* the Tower gives to Paris.[67]

A Corpus of Intelligent Forms

This is remarkable, and we can clearly begin to see what Serres means when he speaks of the hallucinations that structuralism can propagate; in Barthes's framework, the tower is an active object, it inaugurates a technical order, a kind of techno-classicism. The terms in which Barthes characterizes this activity are those of the theological practices of rite and sacrifice; through the tower, "Paris offer[s] itself" to the tourist "as an object virtually *prepared.*"[68] He radically inverts the classical meaning of form as that which provides intelligibility. Rather, elementary forms themselves become intelligible; "this, moreover, is the meaning which we can give today to the word *structure:* a corpus of intelligent forms."[69] This is why Barthes calls the Eiffel Tower a "temple of Science"[70] yet immediately hedges: "this is only a metaphor; as a matter of fact, the Tower is nothing, it achieves a kind of zero degree of the monument."[71] He elaborates, "it participates in no rite, in no cult, not even in Art; you cannot visit the Tower as a museum: there is nothing to see *inside* the Tower."[72] To visit the tower, "is to enter into contact not with a historical Sacred, as is the case for the majority of monuments, but rather with a new Nature, that of human space."[73] What Barthes is contemplating is not so much the technicality of myth, but the mythicality of technics *confined within* a kind of temporal horizon that is neither that of historical myths nor that of the timeless reversibility of the classical order in science and in metaphysics. This confinement is cryptic, but the crypt is profaned within what he calls "the universal language of travel";[74] an order of tourism, that is, belonging to those who are *touring* rather than *settling*, far away from their homes. "This activity of the mind, conveyed by the tourist's modest glance, has a name: decipherment."[75] The panorama that the tower grants the visitor engages his mind "in a certain struggle"—for "it seeks to be deciphered, we must find *signs* within it, a familiarity proceeding from history and from myth; this is why a panorama can never be consumed as an art, the aesthetic interest of a painting ceasing once we try to *recognise* in it particular points derived from our knowledge."[76] And this collective signifying palimpsest in which each plays their part is the great mythic function through which the tower *produces* Paris as a modern city, every day and every moment, anew. The city then becomes "an intimacy"[77] whose connections the tourist deciphers: pleasure, materiality, business, commerce, knowledge, study, habitation. Participating in this function is what "makes every city into a living being."[78] The tower in relation to the city is conceived as neither the brain nor as an organ. Rather, it is "situated a little apart from its vital zones," and there, "the Tower is merely the witness, the gaze which discreetly fixes, with its slender signal, the whole structure—geographical, historical, and social—of Paris space."[79]

The Technical Order of an Object That Is Comfortable

The object without use, the comfortable object, the object-cipher that allows us to experience nothingness—from outside, inside, in the way such as *no thing's details* unfold—this very object is understood by Barthes as a *witness*. We can see now that the tower is celebrated by Barthes as a means to *save* the city: he reveals the presence of the tower in Paris as an object that challenges by being comfortable, through which humanity, by consuming itself, transforms into a new nature. It is indeed a sacred function that he attributes to this performed modern myth. Just as the Greek temples in ancient cities offered spaces where everyone is welcome on sole condition that they submit to the rules of the rite, the Eiffel Tower a sacred space in this sense; "a temple of science," even if only metaphorically so. Every foreigner pays his "initiational tribute," as Barthes puts it,[80] because the site of the tower (as a witness whose survival will guarantee the healthy, non-corrupted, identity of Paris by means of it as a *modern* city) demands, as Barthes calls it, "a rite of inclusion."[81] Everyone must "sacrifice to the Tower"[82]—with one exception, of course; otherwise, it would hardly be a temple: the Parisian is spared. Here, Barthes's mythical structuralism evokes a kind of modern human race: "the Tower is indeed the site that allows one to be incorporated into a race, and when it regards Paris, it is the very essence of the capital it gathers up and proffers to the foreigner who has paid to it his initiational tribute."[83]

I dare to let this stand as it is. Barthes's story, with this aspiration toward modern mankind as a purified race, illustrates many aspects of what is, undoubtedly for many people today, either the source of a predominant mistrust against adopting a structural point of view, or even that of an outright sense of existential distrust/nausea.

Barthes illustrates well, without stripping the stakes from either angle, the intellectual fascination as well as a certain misology (or even an outright intellectual nihilism or fascism) that often tends to accompany trust as well as mistrust. "Being enclosed is … a function of the rite," says Barthes.[84] But,

> How can you be enclosed within emptiness, how can you visit a line? Yet incontestably the Tower is visited: we linger within it, before using it as an observatory. What is happening? What becomes of the great exploratory function of the inside when it is applied to this empty and depthless monument which might be said to consist entirely of an exterior substance?[85]

These words not only characterize Barthes's concern with the Eiffel Tower as a modern myth, but also can easily be applied to the problems of totalities and totalitarianism at large. Barthes's proposal, let us summarize now, is to *affirm* as bare of all meaning the truth of an inclusive total—as a cipher for every meaning—and to hand over all questions of balancing to the domain of sense and signification, in a play of deferred referentiality that can never in principle be bound to "add up" at all.

How to Reason the Sum Total of All Archetypes?

With Serres, we have seen, this is different. There are "real dark meanings" which, he wants to affirm, can be turned into the object of rational, formal, science (and which are not thereby bound back to the administration within regimes of sense and signification). The crucial point is that for Serres, structures are not to be treated as forms that would *contain*, that mediate insides to outsides, and that need be nested in a hierarchical order amongst each other. Forms are to be regarded as *conductive*. The work they do is "to reconcile truth and sense"[86]—for this, they must be void of meaning. For Barthes, we can easily see, the cipher building of the Eiffel Tower does have a meaning: it embodies the meaning of nothing. This is how Barthes can say that the forms of which the tower is built are not only intelligible forms but also are themselves *intelligent*: their performance is that of a distributive, incompletable inclusivity. Serres's approach is, once again, somewhat different. Barthes is fascinated with how the accelerated self-referentiality of different intelligibilities is capable of lending a kind of material presence to that which is immaterial, incorporeal (the tower manifesting "materially," "a corpus of intelligent forms," "the category of concrete abstraction"). The panoramic view of Paris mediated by the tower "lets itself be affected by a kind of *spontaneous* anamnesis," Barthes says.[87] Due to this, the forms for him are intelligent, triggering a kind of spontaneous memory of the immemorable. If forms, after all, are to provide intelligibility, they can only be considered intelligent if among themselves, they are *not* of an equal order. Otherwise, they would be transparent each to the other. How then could they trigger "spontaneous anamnesis"? This spontaneous act of intellection is conveyed in Barthes's phrasing: "it is duration itself which becomes panoramic."[88]

Now let us ask—as we might when trying to solve a Sudoku or any other formal puzzle—what would be the inverse to "these givens"? What would be the inverse to an understanding of duration rendered panoramic? If Barthes is fascinated by the way in which different intelligibilities are capable of lending a kind of material presence for the immaterial, Serres on the other hand is fascinated by the way in which something material (a dark singular content) is capable of adopting an immaterial presence that can be grasped, a graspable presence that does not exhaust this singular content's darkness, unsusceptible to any overall appropriating elucidation. Serres's question is not how it is that all singularities can be inclusively integrated by the work of one— homogeneous albeit mythical function ("The Tower is indeed the site which allows one to be incorporated into a race"[89]). Serres's question is: How can singularities be reproduced in order to engender a pluralism of heterogeneous complexes, which, nevertheless, can again be reasoned in terms of their singularity, without producing contradictions? The crucial point is that Serres is willing to let go of the logical criterion of *exhaustiveness*. For Barthes, this is not an option: the Eiffel Tower is of a "technical order."[90] The sense of intuition he is interested in developing subordinates forms and their mathematical aspect to a kind of reading ("it gives us the world to *read* and not only to perceive";[91] "it corresponds to a new sensibility of vision"[92]) that is mythical in that it "has provisions" but "no utility or use." For Barthes, its functions cannot be rationalized—they can be interpreted only relative to an individual's subjective faculties. The architectural utopia Barthes pursues speaks to the tower's absolutism, its *autarchy* ("The Tower can live on itself"[93]). For Serres too, the functions of dark meanings cannot be rationalized absolutely, but only relative to degrees of access to dark contents within.

Having considered their proximities and similarities, let us try to maximize now the contrast between these two positions: where Barthes opts for *autarchy* within a political state of a modern race that he seeks to evoke and to witness in its becoming, Serres opts, with his white metaphysics and his exodic knowledge for a *civic anarchy* with regard to a natural rather than political philosophy. The objective sense of intuition Serres promotes is also a sense of vision, but not predicated on duration as a panorama. Serres's sense of intuition is attributed to the objective, impersonal agency provided by theoretical geometry; geometry, as we have seen, that serves to stop time projectively in numerous ways, as homothetical spaces that can be bridged and reconciled in their neighboring coexistence with increased degrees of abstraction (more richly differentiated scalarities). Here resides the true scope of Serres's approach to communication via the domain of the "instructable third."[94] For Barthes, the determinism of a materialist philosophy can be kept in check by subordinating matter to a regime of time, whereby "spontaneous anamnesis" (the mythical function of the tower's intelligent forms) provides for the spacing out of one moment into duration, as panorama. For Serres, mythicality does not provide the means for the speculation of durations. Mythicality applies not to the form, as it does for Barthes, but to the content of what needs to be reasoned: there are meanings that are "essentially dark," he maintains.[95] For Serres, the determinism of materialist philosophy needs to be rationalized through the means of physics, not myth. Anamnesis, for Serres, is not singular but plural, and it is never spontaneous but acquired, through exercising; anamneses are achieved through "thinking" (daringly, mystically) *in* mathematics, and through relating the demonstrations to the modeling mathematics facilitates, a kind of modeling in which anamneses are tangled up with sophistication. This is crucial for Serres's notion of structure and structuralism, which is at the core of his realist classicism. "When we pose the question of truth the sole line of guidance we have is mathematics. When we pose the question of experience, what we are left with is mechanics, physics, or natural philosophy,"[96] this is our epistemic condition for Serres. With regard to matter, he maintains, "Matter remains an empty metaphysical word, with neither value nor foundation in the physical sciences."[97] Furthermore, "mass is basic, as fundamental as space and time. Physics knows it, since it makes them its first three dimensions: traditional metaphysics only knows two out of three of them."[98] Where Barthes puts deciphering as the reading of signification primary, Serres places deciphering into orientation with a physics that deals rigorously (meaning quantitatively) with mass—that is, quantum physics. As discussed at length in Chapter 2, Serres's understanding of communication through information theory, is centrally informed by a quantum physical notion of mass. For now, let us develop this point briefly by seeing how Serres both follows and branches off from his teacher, Bachelard. Bachelard's own postulation of a neoclassicism, his *New Scientific Spirit*, took the form of a psychoanalysis of objective insight.[99] In this Bachelard was for Serres, "the last symbolical analyst" in twentieth-century philosophy of science.[100] Let us see in what way.

Toward Critique with Regard to the Symbolic Alchemy of Mythmaking

Serres pursues the question of the meaning of cultures with regard to the central roles of mathematics and physics for his realist classical thought subsequent to the passage cited earlier; "all we have is the sum of all archetypes which the memory of humans passes on since time immemorial."[101] Let us reconsider more closely now what Serres calls an "archetype." Ever since the idealized order of the classicists in seventeenth-century philosophy, the mathematical notion of order understood as the order of *pure* forms, adopted a certain tyranny with regard to the novel domains of the *application* of mathematics. As we saw, the response of the Romantics to this was to declare ideal order a myth, and to adopt the making of myth as its very own method in philosophy. Since then, Serres maintains, "every methodical or critical question revolves around sense, and, if I dare to say so, around its quantification."[102]

So how does this work, exactly? Take any form "that you might want to couple with a methodical function," Serres suggests. "Let's assume that we want it to make sense, that we want to endow it with signification."[103] We can successfully do this when "we overburden it" with material, historical, human, existential meaning. We keep doing so "until it turns into a singularity."[104] Then, Serres tells us as if we all aspire to be alchemists, "this form turns into an archetype." An archetype so constructed is, "the basic ground of reference for a symbolical analysis."[105] If we think along these lines, with Serres, the fact that Barthes has to refer to an architectural (not architectonic!) object in order to postulate that the only truly modern approach resides in the order of readings is not surprising. Serres continues, "The language of sense has only archetypical expressions. It speaks only in ideograms."[106] Furthermore, "Oedipus—a proper name that has become a common name ('un nom commun')—is an ideogram. It enables us to speak the language without a language of the unconscious."[107] Just as with the Eiffel Tower in Barthes's account, which—disturbingly—is explicitly meant to engender a new race, Oedipus is also placed at the beginning of a certain understanding of human nature. In the language of sense that symbolical analysis is engaged with, it is not possible to express oneself with letters which are, regarding their content and their possible relations, indefinite.[108] Symbolical analysis does not deal with letters (characters of an alphabet) but with ideograms, because it is only possible to craft "synthetic paintings" with it, "overly charged images." The more a form adopts a symbolic character, the less can it be treated formally: "The archetype is the maximum in significant overdetermination, be it god, heroes, or the elements."[109]

This latter specification, that the archetype is the maximum in overdetermination of sense—"be it god, heroes, or the elements"—gives us the crucial index to the relation between Serres and Bachelard. Bachelard employed that same kind of symbolic alchemy, Serres maintains; overdetermining a form until it becomes a singularity, and then symbolizing it by recourse to myth. But Bachelard conferred not historical mythical names for those ideogrammatical domains of symbolic reason; rather he adopted those of the four elements of nature itself: fire, water, earth, and air rather than Apollo, Dionysus, Zarathustra, Oedipus, or Elektra. With this gesture, Serres argues, Bachelard must be regarded as the last symbolical analyst. Why the last? Because the myth of the four elements is not merely another myth of an origin—it is the origin of myth itself: "No myth precedes the myth of the four elements," and "the constitution of the world goes ahead of the originality of history. Every mythology is subordinate to a cosmogenesis."[110] Bachelard designated the mythical ground of the four elements as the deepest reference from which his otherwise entirely scientific models are drawn. He endowed the natural elements with a "meaningful, material, power of imagination."[111] The symbolic analysis Bachelard proposed as the beginning of a new neoclassicism, creates "in a dazzling short circuit" this reference ground for distinction, for choosing archetypal models.[112] Why is this a short circuit? Because Bachelard did not think of himself as continuing symbolic analysis. His ambition was to find a way in which cultural contents would finally no longer have to be neglected by the formal analysis of the classical sciences. His entire project of giving infinite insight into what can be objectively depicted, mentally, and theoretically, his entire idea in short of developing a psychoanalysis of imagination as an objective, material force, depended—in order to truly count as a formal method—on a key principle: the principle that in order to "go from symbolism to formalism, in order to go from a notion of the model as goal of the method to the model as problem, one has to verify that the variation of the basic sum, the matrix or basin from which the archetypes are sourced, is exhausted."[113] In order to determine such exhaustiveness formally in logics, it must be possible to assume an extension of the domain at stake. For Bachelard's designated elements, as the entirety of nature, this criterion is not met: since this is "nature at large," it is not possible to conceive any extension. Nature, in the scientific mind, is what does not admit for anything else more primary.[114]

A Realist Classicism

And yet, to claim a formal method that need not determine in terms of exhaustiveness— is this not what we have just stipulated for Serres as well, in distinction to Barthes?

Yes and no. We are coming closer to exposing and delineating an abstract situation general enough to return to the term "structure" as "an architectonic form" and as "a precise methodical concept" in Serres's framework. After Bachelard, according to Serres, the only thing that remains to be done for philosophy to open up a new classicism is to develop a notion of *criticality* (not *critique*) along structuralist terms. Such criticality promotes analysis focused on the projective inverse domains to those of symbolical analysis. Such criticality consists in voiding form from all possible sense or signification. It consists in *thinking of form itself formally*: in thinking of a notion of form that casts off, is delivered, born, isolated from all meaning—this, in short, is what Serres understands by structure. In his philosophy of the transcendental objective, structure is regarded as an operational form, a form that is performative in the quasi-domain of the projective, and therein it is objectively reasonable. Structures give us spaces of geometrical parallelisms that are called, in mathematics, spaces of *homothesis* (see Chapter 4). This quasi-domain is objective and yet transcendental, in the sense that it can never itself exhaustively be rationalized; but this transcendentality is also objective in the sense that it is based in things themselves, not in the faculties of a human subject alone. All things communicate, in Serres notion of nature; this is indeed why he proposes a natural rather than merely a social contract in his perhaps most prophetic book with respect to the Anthropocene, *The Natural Contract* from 1992. The categorical status of this objective transcendentality is that of the "quasi." Things themselves are all to be addressed, we have seen, through the sixfold panonyme of the world: *Pantope* (all of its places), *Panchrone* (all of its durations), *Panurge* (the universal worker), *Panglosse* (all of the spoken tongues), *Pangnose* (all of knowledge), *Panthrope* (all sexes, instead of only man as in "anthropos"). Things themselves can only be modeled structurally, Serres's realist classicism maintains, because the proper name of the world is conceived as a group concept (in the mathematical sense of "group," which gave rise to a formal notion of structure in algebra, subsequent to Evariste Galois and Nils Henrik Abel in the eighteenth century). The summoning of this polynomic proper name for addressing the whole world as one does not seek to yield the depiction (and facilitate the recognition) of one grand panoramic duration. Rather, it seeks to keep categorically distinct all of duration, all of places, all of knowledges, all of languages, all of sexes and all of work that is not only *public* (*demiurge*, as it is, arguably, for Barthes) but also *universal* (*panurge*) by dint of subjection to any one hierarchic order among them in particular. It is the proper name of the world, to Serres, and as such it is of a metaphysical status: the categorization it facilitates is that of the domain of the pseudos or quasi—to which Hénaff referred with "the games of the logoi." In Serres's realist classicism, the *plain of Aletheia* can be addressed as the locus where exodic reasoning withholds the power of the pseudos. It turns into the architectonic site of *objective* anamneses and *objective* sophistication. We have discussed the impersonal cogito at work in such objectivity as the *logiciél intra-matériel,* an intra-material kind of software. The domain of the quasi is conceived by Serres as transcendentality relative to objects, their capacity and abledness in doing what they do according to Serres, namely providing for recognition and regularity of course (formality), but also always bearing traces, behaving as memories, and banking time, as we saw in Chapter 5. Form originates in death, for Serres; we can think of his intra-material software as something like the spectral quickness proper to form from the point of view of physics thought about through the lens of information theory. What we will try to grasp better now is the resistance of the domain of the quasi so conceived to any simply social order—a resistance that is in place unless social order is accompanied by some reintroduction of the element of the sacred (the pure). As we have seen in Chapter 6, "The Incandescent Paraclete: Tables of Plenty," Serres does not hesitate to affirm a certain "holiness" within his conception of nature via an intra-material software, but he entirely disinvests such "holiness" of the sacred. This is perhaps the most eminent difference between his conception of the quasi as a metaphysical domain and what Bruno Latour terms "the quasi-object." The ideas rest on different presuppositions.[115]

Objects then are not "free," it literally *matters* how they articulate form. To assume a transcendentality relative to objects, this places objectivity in a contagious zone of purity and impurity. But to conceive of such transcendentality with respect to experiential domains that must count as metaphysical (rather than simply empirical) furthermore entails that objectivity is not only contextual and situational, "subjected" (by limited capacities and abledness of objects), it also entails that the limitedness at stake is *principled* in a universal way. The metaphysical status of a domain of the quasi in consequence provides for the irreducible mediation between the real (situational, local) and the rational (global, general); we have seen how Serres conceives of this relation as one of equipollence (the real and the rational ought to be regarded as equals in force, validity, or power) that does not subject the real to the rational or the rational to the real. Such a notion of formality is inevitably cyclical, conductively circuitous: a *model* that is to *realize* an order for each equality means that the identity relation at stake needs to be worked out and contracted situationally, and restlessly. Because the three models that must together and in parallel "principle" (govern) this relation inevitably produce conflicts between them (we can think of power as relating to politics, force to nature, validity to ethics). To think of formality as being inevitably cyclical, conductively circuitous, is no mere fancy of an abstract intellectual athleticism. It is at the root of the key assumption from which Serres's architectonic approach to information and mathematics departs; namely, that form is not to be set apart from a particular content (real and dark meaning) that it could grasp; this would be the gesture of introducing the sacred (that which is set apart, purified). Rather form is to be distinguished against the generic background noise—*le bruit de fond*, the noise of the ground, the confluence of different geneses, in short the confusion of a *principled reason* whose principle it is to be indifferent to the representation by any one particular *candidate* (to step in the vicarious position of the principle). It is a principality that serves a natural philosophy, not a political one, and that introduces a way of addressing subjectivity within nature and as nature, that means: objectively, as object among objects. Form, as well as ground, have to be addressed not through a notion of order directly, but mediated through the reciprocal and mutually relative *measures* of order; in short, positive and negative *entropy* have to be considered with respect to both energy and information.

"History is the island of negentropy in the entropy of culture," as Serres puts it.[116] This is the crucial formula with which he sets out to show the way towards a realist classicism. Analysis in these terms is no longer merely symbolical nor merely formal: a critical approach that proceeds structurally renders analysis *communicational*. The formality of such analysis is cyclical and conductively circuitous. Its symbols as well as its forms are now always already modeled in terms of information theory, or in other words, in the equatorially polar terminals of symmetry structures that conserve order invariantly—order that can be addressed and exposed through models that realize it, in a great variety of manners and on many heterogeneous scales.

The objects of structural analysis are no longer to be described simply as more or less pure, or more or less faithful to their mythical origin. They can now be addressed as quasi-objects, objects that are not only in circulation but also constituted circuitously. They are objects that are formally distinct, but nevertheless *mass-ive*, actively rumbling with noise that threatens to annihilate them, but that also makes them what they are (in their particular ways of behaving as memories). Quasi-objects demand that we develop, once again, humility and a respect for ideas as a puissance. The reconciliation between truth and sense depends upon developing a transcendental sense of intuition that manifests objectively, and that renders us capable of what Serres calls "historical doubt."[117] Formulating the form of transcendental objectives is what the complicated wording of "models of transparent structures" are all about. Those models support not so much the reasoning of historians or critical theorists, but a new synthesis of both; *critical historians*—what Serres calls later, in *Les origines de la géometrie* (1993), scientists of history.

Scientists of history commit themselves to an exodic (rather than a methodic) discourse. An exodic discourse is one where the epistemological maxim for all methods, namely, that the path demanding least expense while promising maximal effect is not a safe guide for objective insight. I want to suggest that we can think of such discourse as discourse that not only *conveys* but also *confers* senses—from the Old French *conférer*, for "to give, to converse, to compare," from the Latin *conferre*, for "to bring together," figuratively "to compare, consult, deliberate, talk over." In short, we can think of exodic discourse as discourse that concerns itself with experience, but it does so by making use of "the power of *pseudos*" (the feigned, the insincere, that which is only apparently what it claims to be) which the sophists had introduced into philosophy. The sophists use the logical forms as a mere means of arguing in their games of *logoi*; thereby, the question of truth—the being of the thing—became arbitrary. Serres too is sensitive to what Hénaff calls "ontological deceit."[118] For him as well it is not the moral falsehood or malice of an individual subject that is at stake, but rather something of an order that is objective. But the objectivity Serres's discourse is concerned with is that of *objective experience,* and it entails metaphysical imports of whichever kind. Exodic discourse is to examine experience's singularity objectively. Where Hénaff's critique of the sophists' games sees ontological deceit, exodic discourse counts, treasures and banks on an abundant past where what looks like betrayal can reveal itself as the manifestation of a transgression—yet not transgression as a moral or legal act, but as an economic act of metaphysical import or export. Like methodic discourse, exodic discourse proceeds by analysis—but analysis is symbolically economic here; it is parasitic, not concerned with purification. Analysis here is structural, it produces models in which every step of differentiation is complemented with an inverse step of dedifferentiation. Such models realize the transparency with which they operate analytically. Structural analysis so conceived is about *instructing* a domain of the quasi, whose architectonic status draws credit from metaphysics (universality). It is an architectonic domain that accommodates the power of the *pseudos* as a *withholding* power, by giving it an public face. Such analysis makes it possible to learn remembering objectively and collectively an abundant past; a culture worthy to be called "universal" can unfold from such a domain. This I understand to be the ambition of a realist classicism.

Metaphysical Import: Architectonic Models That Conserve an Analogy

The full title of the article in which Serres maintains that "our time appears capable of re-conciliating truth and sense" is "Structure et importation: des mathématiques aux mythes." This reconciliation depends, as we have seen, on the capacity of "structure" to facilitate the "transfer" of concepts (that conserve dark contents) from one field to another. The article was written in 1961, during the rise of what later came to be called structuralism; the question at stake for Serres was if and how a concept like that of *structure* could be reasonably transferred from the quantitative and formal theories that apply them in the natural sciences to the qualitative and semiotic theories of the sciences that study history and culture. The claim of such possible reconciliation between truth and sense rests on Serres's attributing this particular notion, "structure," the capacity to deal quantitatively with the cultural contents of myth, and this without appropriating and dominating (identifying through representation) its dark content. Structural analysis is capable of formalizing particular myths in their variability; but what is of key importance is that structural analysis, as Serres introduces it, is capable of *inventing the mathematical inverse* that corresponds to the symbolically analytical ways of formalizing myths. This is how his structuralism can be a form of critique. The acquired mathematical inverse then provides spaces of placeholder positions; vicarious spaces, where a plurality of cultural meanings (variations of a particular myth) can be transferred into a third, architectonic, domain that is capable of accommodating great varieties of those cultural meanings and allow them to coexist: inventing such formal notions of spatial domains is how the mathematical notion of structure can facilitate, by drawing credit from what Serres calls *white metaphysics*, the import of concepts from one field to another—even if the contents of those concepts lack a fully rational basis, even if those contents are what Serres calls "the dark contents of a Dionysian world," the contents of experiences in their singularity, in which "the human soul, emotions and the fate of mankind find expression."[119]

Such a notion of structure as providing formalized webs of positions in terms of vicarious placeholder positions is key to how structural analysis can facilitate the *importation* of ways of observation and methods of analysis from one field to another: "What we call importation is this: a methodological concept, which is clearly and precisely defined in one determined domain, and which was successful (methods can and ought to be evaluated only with regard to the fruits they yield), be tried out ad libitum in other domains of knowledge, of critique, etc."[120] The very possibility of such import is entirely dependent upon the following: "To remain clear and precise, it's necessary to avoid any deviation and any ambiguity when importing the idea of structure from polymath theories [*des théories savantes en général*] to the field of cultural critique."[121] A polymath idea of structure has been formulated, and formalized, in algebra: "In algebra, the idea of structure has nothing mysterious about it."[122] In short, as Serres puts it, the import of concepts is possible if we can build a machine for how a concept *operates*,[123] that is, when a concept is highly formalized, when it is possible to construct a machine in analogue manner to a particular domain that is systematically analyzed.[124] Serres is well aware that this sounds like a scandalous proposal to many: "To those who find this scandalous we point out that we have already proposed machines that work *like* Darwin's system. The idea is not new. It appears abominable only to those who despise machines, and who don't know what they are, what they can be, and must be."[125] A structure incorporates virtually, in the logistic order of a vicarious space in which an action is unfolding, the concept of what we could call *an architectonics of mobilized states.*

If we pick up (and adapt) the famous phrasing by Immanuel Kant, who spoke of architectonics as the art of building systems, our suggestion here is to speak of an architectonics that would be the art of building active states in which nothing is fixed and all is mobile. Serres's *principality of reason*, in which no particular principle is to be representing the *official* order in an unchallengeable way, confronts us with a sort of anti-architectonics (if architectonics is meant in the Enlightenment sense of Kant): it confronts us with an architectonics whose building craft consists in "un-building" (undoing) rather than in "building" (doing). It is an architectonics of the natural domain of a withholding power, that manifests in the articulation of a quasi. But is the result of such reason, is knowledge in the terms of such architecture not quite simply a great accelerator for the production of entropic dissolution? We will come back to this in the "Coda" to the book. For now, let us try to stay literal, when using this metaphor of a great accelerator in nuclear physics like that in CERN. These accelerators are observatories. This is how I think we should think of Serres's architectonics: as an observatory for scientists of history.

Can we think in a generalized sense of such a notion of models (models that do not represent but that realize, relative to an architectonic domain from which they draw the credit for metaphysical import) as *observatories*? Are all models observatories, when conceived as machines whose purpose is to be communicative in the sense of a symbolic economy (rather than being productive, as in a lowly sense of economy)? Are all models that preserve an analogy, for example, *models that work like the cosmos* in the case of the CERN accelerator, or models that work like Darwinian evolution, to pick up Serres example, *observatories*? We must come back with this to the difficult distinction between the sacred and holy at this point. The domain of the quasi is a domain of transcendentality that applies to objects, and that renders objectivity "subjective" (in the impersonal manner of the "it") within orders of objects among objects, we said. There is a holiness to this transcendentality, we argued, but no sacrality. We can see now better how this is so: models so conceived (as preserving an analogy, as acting as observatories) are always relative to a *module*. A module constitutes a one-over-many relation, in algebra as well as in architecture it is literally "an allotted measure." Such modules *constitute* particular orders in that they incorporate a *base magnitude* from which all proportions of a building are extracted. With regard to them, a built structure can be consistent, systematic, deductive *in how it is being designed*. A module provides a reference ground that supports well-proportionable articulations of a particular invariant order. As we can learn from the history of architecture, such modules are quite literally *symbolic pedestals* (classically, they are the base of columns). I want to foreground architectonic orders to emphasize an important difference of their systematicity to the notion of system in logics: while the latter strives to be as comprehensive as possible, architectonic orders are always *reduced* models. They are models meant to "realize" (not "represent") the order to which they contribute as integral parts. The reference "order-ality" of architectonic orders always coexists within several such reduced models.[126] Let us recall from the preface to this book that Serres thinks of his own epistemology in the terms of a general treaty on sculpture. This relates to his view according to which form derives from death. In his second book on foundations entitled *Statues* (1987/2015), Serres suggests conceiving of such modularity like we conceive of the various performative rituals of dealing with death.[127] The base for a reasoning in terms of a totality of archetypes, and a totality of world views, he maintains thereby, must be approached within the registers of anthropology. We saw earlier that a key concern for Serres is to refuse the sophists' maxim that man be the measure of all order; rituals of dealing with death are human, but Serres finds models in them—and this is where I want to direct this discussion of an architectonics of knowledge to—of *how to preserve the analogy to what transcends human measure*, namely, death.

As Serres develops it in his foundational trilogy, anthropology has a say with respect to the origins of geometry. But having a say is not the same as dominating; if dealing with death would not be the analogy preserved by the various rituals, if a particular ritual would step up and claim to represent our relations to death uniformly, then the broken link between holiness and sacrality that is so important to Serres would be "fixed." Then, questions of form and geometry would be dominated by anthropology. Especially in the third of these books, entitled *Les origines de la géométrie* (1993/2017), Serres develops how geometry ought to be thought of as a confluence of multiple geneses, whereby the origins that he explicitly addresses are the following: juridical, political, discursive ones with regard to customs and laws; and optical, ethical, astronomical, algorithmic and automatic ones with regard to nature. Geometry's originality never ceases to be contemporary, because it results from a confluence of geneses. Geometry is pure in that it is impure, just like reason is principled by subjecting to no one particular principle. These apparently paradoxical phrases are *not* sophistic articulations—in a reason that proceeds by architectonic modeling they relate *paradoxicality* (sense, doxa) with *absurdity* (reason, grounds) in a fashion in which they are both subjected to each other. Like this, they capture and express inversive thinking that is rational, and that relates to real experiences, not to an ideal representation thereof. These formulations are *historic* in the sense that they capture and express inversive thinking as time passes massively, as developed in Chapter 4. Reason that proceeds architectonically by inversive and structural analysis ought to derive its modules not from an ideal notion of order but from nothing but experience.[128] Forms, the bearers of rational regularity, originate in death, he maintains.

When Serres maintains that those "dark meanings" of the Dionysian world, in which "the human soul, emotions and the fate of mankind find expression"[129] can be studied scientifically, Serres asks us to accept that there is an *essential* darkness to meanings, yet one which does not contradict clarity and lucidity. Darkness and whiteness together form the polarity of a quantum-physical notion of "massive lucidity"—a polarity that must count as constitutive for the technical forms of spectra (Chapter 2, "Quantum Literacy"). Meanings are *in their essence* impure, confused, complex, sortable in a thousand ways. Essence corresponds to universality. This is not a link which Serres sets out to break. If meanings were not universal, if they would not claim a timeless actuality (what we called, with regard to Serres, "contemporaneity in the agedness of the universe," Chapter 3), we would not speak of meaning at all. We would speak of *things appearing significant*, in one way or another, for some one or another, in one moment or another.

What changes in Serres's formulation is that those meanings are not supposed to be made the object of an individual subject's light-bringing reasoning; rather, the lucidity is considered to shine forth from an intelligible order that is impersonal and objective. The subject in Serres's experience-based realist philosophy does not seek to dominate those dark meanings; rather, it seeks to be instructed by them through learning how to model them structurally: the subject can learn from the darkness of objects how to live humanely, socially. The subject learns to master those dark meanings in a manner that needs not violently dominate them—like Thales, at the foot of the Pyramid, did not have to violate the immenseness of the monument's dark meaning. Geometry facilitates the invention of means to take measure in ways that combine precision with finesse. Its pride is in training a form of mastership that excels in learning how *not* to take things into possession. It establishes a world of objects among objects.

8

Coda: Architecture in the Meteora

We had thought we'd die from totalization; here it is that we can perish from splitting up. Everything happens as though violence was equitably dividing up its ravages. Might it be universal like geometry?

Michel Serres, *Geometry*[1]

Let us collect the different strands of how to engage with time as developed throughout the book, and see how one might engender from it a corresponding understanding not only of architectonic reasoning, but also of architecture in a (humane and historic) world of objects among objects. Is there a way to think of architecture in the meteora? Let us remember where this interest comes from. Serres is ready to grant that all things may be miracles. But if we want to reason the experiences they trigger, we cannot accept their miraculousness as an explanation. Instead, he maintains, we should ask with respect to anything at all that troubles our understanding: "If it is not a miracle, then can we build it?"

Architecture in the meteora could not be architecture "at home." Let us begin by asking for whom it would be building. We live today in an age of unprecedented "progress" and "acceleration." The wealth in our world is enormous, but it spreads with great inequality, and it also produces great harm. Our current philosophical discourses are witnessing this everywhere, and often refer to it through the notion of the Anthropocene. Is not Serres's affirmation of providing a philosophical discourse that "welcomes us at tables of abundant plenty" (as we have argued in this book), and that helps us "to deal with a lot" by affirming chance as objective—as a philosophical principle, even—ultimately about proclaiming a naïve and dangerous optimism that despite this growing inequality, despite of the harm we are inflicting upon the environment, all is and will be, in principle, "well"? This concern cannot be refuted polemically, even though what is at stake is entirely political. Here is a passage from *The Incandescent* that is relevant with regard to Serres's epistemic optimism.

In going back up ordinary evolutionary time by dedifferentiation, we [human kind] returned, if I may, backwards and went from the many species, well named since specialized, to a kind of common genus. Non-specialized, humanity became, if I may, a counter-species: literally, it became generalized. Losing its specifying characteristics, it planed down its programme and became a generality. Humanity, that unknown: x with every value because having none.

Becoming human tends towards this white indetermination. Zerovalent, omnivalent, nil-potent, totipotent; good-for-nothing, good for everything. Every bit of progress, stroke of genius, invention or discovery originates in such a backwards movement and advances by choosing from among the range of a totality opened in this way. Consequently, human nature or, if you like, human nascence [*naître*] can be defined, without definition, as a tendency towards this forgetfulness, this deprogramming, this dedifference. Who are we? Indifferents. I exist and think in a point where nothing concerns me.[2]

To say that "I think where nothing *concerns* me," this precisely does *not* mean that reason has built a house where one depends on nothing and lives in an autarky. Quite the opposite, as we have argued throughout this book. This passage speaks of a place where the cogito is *contingent* with everything else; this is the cogito of the impersonal "it" that manifests physically in all existing things (insofar as everything sends, receives, stores, processes information, organizes negentropy and produces entropy). If Serres's philosophical discourse is neither about proclaiming naïve optimism nor a frightened conservatism, then what could possibly be its concrete outlook, what could give us direction (sense) if we submitted to its proposals?

Let us come back now to the point in the last chapter where we described the sophist's power of the pseudos as a power of withholding, and proposed an architectonic domain of the quasi (as an objective transcendentality set relative to metaphysics, not to empirical science) that pertains to the projective conception of the world-in-general; a world of objects among objects. We described the projectivity behind such a conception of a world-in-general as architectonics whose building craft consists in "un-building" (undoing) rather than in "building" (doing). Is the domain of the quasi, viewed like this, not quite simply a great accelerator for the production of entropic dissolution, we asked. Our preliminary answer was affirmative, but it was only half of the answer. We suggested to take this metaphor of a great accelerator in nuclear physics, like that in the CERN, Geneva, in the sense of a *science of literalness*, as we proposed in relation to Hénaff's approach to the problem of the price of truth. These accelerators are observatories, and our suggestion was to think of Serres's domain of the quasi as an observatory for scientists of history. So yes, de facto they produce entropy; but what do they "produce" as observatories? What kind of theoretical insights and outlook are they capable of engendering? What Serres's discourse can help us to see is that with the technical mastery of quantum physics we have an architectonic of reason that operates according to a novel kind of mechanics of light, where philosophically speaking, light is the opposite of what our Enlightenment traditions thought it was: light is not "immaterial," it is "substantial." With such a conception of light, and the mastery of its mechanics, we quite literally have new optical instruments today. This affects the light in which we see everything. Quantitative science can now attend to "soft" contents as much as to "hard" ones, and "hard" ones acquire a certain contingency which had been reserved for "soft" disciplines only. From such a viewpoint, science at large is turning through its current form of industrial specialization into architectonic dedifferentiation, manifesting in the digitization and datafication of our world's infrastructures. This raises the problem that Serres never got tired of foregrounding: If we consider the domain of the quasi as a "domain," if we view it with regard to the "domesticity" it can provide, this process of dedifferentiation totalizes everything within an economy with neither orientation nor history nor politics. An economy does not make decisions, it processualizes decisions with the sole interest of providing for "sustainability" (let us refuse, here, the misologist "diagnosis" that would put individual "profit" here). The more general the order of such an economy, the more entropic its effects. In its most general forms, economy keeps things in circulation at the cost of dissolving the historical integrity of specific facts, events, structures, and cultures. An economy is indeed a machine producing entropy (now also specifically in the sense of pollution and waste). But only on its most general scale is it also a great accelerator. This is a result of thinking about the "progress" in mastering such architectonic reason in the registers of industrialization and production. It looks different when we think of such mastery in terms of a physical communication paradigm. We tend to forget that the machine is a "built" machine, conceived with architectonic reason that is at work with mass-ive communication, by handling it as a quasi-material, with a novel *mechanics of light.*

With respect to such a two-way notion of "building," we must ask now also the complementary question: Can the machines of such architectonic reason be set to work inversely, and articulate (build, design) negentropic forms of organization, according to the generalized equatoriality we proposed earlier? We need to look for a second pole, inverse to that manifested by the particular machine of the great accelerator.

Where could we possibly find such a second pole? Nuclear physics is a form of technical mastery of nature, and it has widely troubled philosophical discourses throughout the twentieth century. But artificial photosynthesis is equally so a form of technical mastery of nature; it is equally abstract, and inverse to nuclear fission (molecular chemistry, photovoltaics). Artificial photovoltaics is of the same scale but inverse to nuclear physics, in that it does not seek to master the concentration and maximization/minimization of energy, as the latter arguably is; its mastery is directed at learning to "conserve" energy according to nature's given wealth in how forms of bodily organization conserve and metabolize what we can call *meteora alloys* of energy and information ("impure particles" of Serres's *logiciél intra-matériel*, mixtures of hardware and software). By taking into account this complementary inverse to the paradigm of nuclear physics, can the architectonic reason that is capable of "building" such accelerators also serve to *differentiate* the forms of observatories our novel mechanics of light is capable of building to an inverse end? Can the same reason be turned as well toward countering the acceleration vector of scientific progress with one of deceleration that is of no lesser puissance, equally scientific, sophisticated, and progressive, yet in counter current, in retrograde manner—inversely so, namely, in putting technology in the service of conservation? The *architectonics* of a domain of the withholding power needs to be complemented with an *architecture* in the meteora.

Let us remember the context of Aristotle's treaty on *meteorologia.* Why did Aristotle write this book, which sits so oddly in between his two other treaties which separate the earth from the heavens, *De Generatione et de Corruptione* and *De Caelo*? The former is a treaty on the continuous transformation of natural material things, while the latter treats the unchanging order of the stars and planets. The *meteorologia* sits so oddly between these two because it treats natural phenomena that undergo "change" but do not transform continuously, and that cannot easily be accommodated within either one of the two spheres. The phenomena it attends to are meteorites, the appearance of colors as in the blue sky; it asks why is the sea water salty and the fresh water is not, for example. In order to think about the cyclical processes behind such unsteady and apparently "changing" (not only transforming) nature, Aristotle makes a case to engage with impure mixtures in their own right. He proposes material "principles" that "count" respectively on different scales of natural phenomena, but always also integral to the most general scales (which are treated in the two other treaties on biological and cosmic nature). Those material "principles" are integral but not reducible to them.[3] An educated guess would be that Aristotle is interested in a theory of optics (and the mechanics of light which provides the foundations of this discipline). Given his commitment to a realist philosophy, the optics he was interested in developing needed to be able to account also of these apparently "insubstantial" phenomena (the "meteora" as mixtures). The Renaissance scholars, who were developing novel theories of optics as well, arguably followed Aristotle in this same realist gesture.[4] This is why a theory of the meteora provides the inverse and decelerating vector capable of complementing the accelerating vector of the "pure" disciplines, those concerned with the most general orders. Arguably, this is also why architecture has always been firmly situated in this sphere, that of the meteora.

Nevertheless, a novel architectonics (novel mechanics of light) will engender a novel architecture—architecture now as the sophisticated refinement, via models that "reduce" distinctly, decidedly, and prospectively the enormous amount of potentiality the novel architectonics unleashes. What would be first preliminaries for an architecture in the anthropocenic meteora? Such an architecture could not be architecture cultivating an imagined "home." Can it be a return to ideas of "building," as in the building of concrete utopia (or to concrete ideas of utopia)? Architecture in the meteora projects the domain of the quasi as a real and feasible locus in quo for the coexistence of as many forms of "commensurate universals" as possible. In other words, it is committed to serve the (scientific, economic) mastery of nature in order to coexist as nature with nature. At stake is a political mastery of nature that governs by striving to let nature "be," to not have to touch and exploit nature. It is a political form of mastery to the degree that science is capable of "copying" nature. Surprisingly, *copia* is the Latin term for "plenty," and *copiare* literally means "to transcribe," "writing in plenty." It means duplicating too, but through this emphasis on the plenty, the relation between "originality and duplication" is no longer one of "closest possible representation," guided by an ideal of rendering the difference between copy and original transparent.

What such mastery is committed to is, in short, to a novel form of politics, not economy. There is a difference between *gérer* (managing) and *gouverner* (steering), as Serres often puts it. The former is the gesture of economy, it is committed to the present and on to how to manage best given the stocks; the latter is the gesture of politics, and it is committed to the long term. For politics, at least four questions are always decisive: where are we (in a particular present moment); where do we come from; where do we want to go; and how do we want to get there.[5] In order to put the contemporary in a long-term perspective, politics needs history. Architecture in the meteora is architecture constituting places within a *Grand Recit* that models the economy of what we called the "quantum domain's domesticity" (Chapter 2). But it cannot think of this domain as a "home," unless it be a universal kind of "home" or "nativity"; within a realist perspective, it ought to think of this domain as the locus in quo where as many different "commensurate universals" can coexist as possible. Such a locus in quo is projective, ideational, despite its status as "realistic"; hence we can perhaps think of it as a form of utopia. Architecture, as against architectonic reason, opens up the possibility for a politics with respect to this alien "domesticity" where no one in particular is native, because everyone is in principle (universal nature), and whose economy is established by the reason that operates with the new mechanics of light. It regards this economy as a symbolic economy. Accordingly, it thinks of the domain of the quasi (the projectivity that models the domain of "quantum domain's domesticity") as driven by a *symbolic libido* that involves "trade" and "transcription" in what we called "metaphysical import/export" where the domestic libido of a "pure" economy sees treason. The symbolic libido circulates *any* world picture, any ideology, the indeterminate sum total of all "dark and mythical contents" that constitute meaning.

An architecture in the meteora would be in the service of politics within such an economy because it contracts power (brought by science to this economy) in the service of freedom. This entails that it gives this power a public face. Let us remember that freedom, for Serres, is a metaphysical notion. It can neither be given (nor taken) by any one social or cultural reality in particular. In *The Incandescent*, Serres gives a short passage with exercises that concern "the unleashed libido" which acts as a driving force in our realities; Serres calls it "the libido of belongingness."

> Does every evil in the world come from belongingness? I'm inclined to think so. Evil prowls in these boundaries, closure and definition, ensuing from comparisons and the rivalries they incite, it being roused by this libido's heat.
>
> May you, once a day, in order to cool it down, forget your culture, your language, your nation, your dwelling place, your village's soccer team, even your sex and your religion, in short, the thickness of your enclosures. Women do change names after marriage; travellers change address, and emigrants passports. Translate [*traduisez*] some foreign word; betray the easy dialect in your mouth; imagine that the person being accused of betrayal is traversing, in the literal sense (traducit), a border, is quite simply travelling and that, whether an importer or exporter, he is giving across this barrier (*trans-dare*). Call this traitor an exchanger instead. Bless the translator [*traducteur*]. Women, marry your brother's enemy. If you live in the shadow of a modest church tower or cathedral, look upon it once a day, at noon for example, as a ziggurat, a pyramid, a mosque or as a shadowless pagoda. Happy religions whose founding narrative doesn't deify their own land but on the contrary blesses an entirely different one, distant, said to be holy and so inaccessible that the land upon which real life unfolds becomes a valley of tears and place of exile. Where are we from? From nowhere, from elsewhere, from

transcendence. Let's lastly practise, during this hour of light, dressing our friend up as a Persian and seeing our dragonesque beasts as beloved princesses calling for help. May we, from time to time, forget our belongingnesses. Our identity will gain from it. With peace, on top of that.[6]

Keeping this libido of belongingness in check, architecture builds, equipped with architectonic's reason *well versed* in the exercises Serres has given us (and similar ones). *Archi-tecture* then does not mean building a first (an original) "home." Instead, it means caring for a place to live in in inverse manner to that of *au-tarky*. Autarky minimizes dependencies and wants to constitute a particular (set of) belonging(s) through achieving the largest possible independence, while architecture maximizes dependencies, by maximizing the contingencies of belongings. Architectonic reason in the service of architecture (with its commitment to distinguished models) thinks of itself as entropic and therefore builds contingently and complexly contracted negentropic orders. In the service of autarky, architectonic reason conceives of itself as negentropic and acts (towards all that is other to it) consumptively, entropically. In the service of autarky, architectonic reason inevitably acts hegemonically; in the service of architecture, it does not inevitably act "good" or "innocent" either; but it can, it acts politically.

With this hint at a novel approach to architecture in the anthropocenic meteora in mind, let us return one last time to Serres's very first book, *Le système de Leibniz et ses modèles mathématiques* (1968). The introduction is entitled *Ensembles théoriques*, and he begins with the subsection *Scénographie, Ichnographie*. Scholars interested in Leibniz share a kind of embarrassment, Serres begins. It concerns the irreconcilability of Leibniz's *rigorously systematic* thought, while this very systematicity does not reveal itself easily to "rigorous" understanding. Leibniz presents his reader, as Serres puts it, with

a potential ordonnance which partially reveals itself and incessantly refuses to do so entirely, a vague idea of a perceived coherence seen a thousand times in oblique, and which hides its *géometral*, the sensation of progressing in a labyrinth of which one holds the thread but has no map. Offered perspectives, multiplied points of view, infinitely iterated possibilities: it never seems that one could actually arrive at the exhaustive limits of a synoptic, spread out, complete and current plan.[7]

What most of the commentaries work out as a flaw in Leibniz's philosophical "system," counts to Serres as the important (and dark) content worth being preserved objectively; that is, not primarily with respect to Leibniz's personal intention or ambition behind writing it, but with respect to it as the manifestation of a cultural content that is given, and that can be received and appreciated through learning its architectonic gesture like we learn the gesture by studying paintings or the play of musicians. Serres finds in Leibniz's philosophy an idea of the system at work which is to comprehend and organize all that obeys the principle of identity where identity itself is substantial, just like freedom is (for Serres). Both are metaphysical concepts, and they govern their subject matters through keeping their own position an open placeholder position—such that no principle in particular can claim it once and for all. The vicariousness of such a notion of systematic order is similar like that which mathematicians see in mathematical order. For them, as we have argued, a formulated new field is not so much an authoritative norm but very much an open field to be further articulated and "built." So is the idea of identity he finds at work in Leibniz's notion of a "system" to Serres. To him, it is the field of a principle whose subject (identity) is to be considered substantial, and conceived through invariance. He assumes nothing more specific for it than that it must be capable of absorbing and integrating all the variations that *might* actually be attributed to it. To Serres, Leibniz looks for a notion of system that could accommodate the possible (not just the variable) and where the possible produces the necessary (rather than being built on it). To Serres as to Leibniz (at least in Serres's reception), such a notion of system can only be found via mathematical *models* and architectonic reasoning.

We find in this early work of Serres already *in nucleus* the discussion we find in his work on the pluralization of anamnesis into anamneses, which we touched upon in the last chapter. Mathematical models are not adequately addressed as representing a reality, whether it is an ideational experience that is at stake or a sensational one; mathematical models are models that *direct* the "realization" of what they formulate and allow to demonstrate. Accordingly, Serres's notion of a philosophical system too, which he uncovers in this early book on Leibniz, is one that lends itself only for mathematical modeling. Mathematics is to give the highest notions (the categories of his metaphysics with its white concepts) but in order to serve rather than to dominate the disciplines. With mathematical models, every discipline can study particular experiences. They lend themselves well for such empirical studies because for mathematical modeling, "there is a plurality of possible paths of deduction."[8] Nevertheless, philosophy is systematic because it is architectonic; mathematical models are "built." Like this, Serres gains a notion of system which is of reversible orders *and* of irreversible ones. The system itself is unitary but "like a scale," made up of a plurality of "orders, derived from an infinitely replicated infinity of infinities."[9] Within such a scalar sheaf of orders, Serres elaborates, "enunciations are universal and they conserve the analogy [a 'discrete multiplicity', which Serres contrasts with a function as 'a continued variation']."[10] This is why mathematical modeling of the system, conceived as a ladder where different orders (each infinite but projectively and scalarly measurable) are leading from one thing to the other, and hence "progressing" but not in a straight line. The "progress" such a notion of system depicts progresses indefinitely. Its steps are governed by "binding laws of one-to-many, finite-to-infinite, which are of value in multivalent manner for perception, liberty, knowing, creativity, remembering etc., which are all at work also in the mathematical model."[11]

With this last quote we can see again clearly why Serres wrote his book in the registers of architectonic terms, those of ichnography and scenography. All those aspects of reality (perception, liberty, consciousness, creation, remembering) feature in every mathematical model for Serres, and yet it is not mathematics that governs them. "There is," he affirms, "no relation of cause and effect here," in this relation between mathematical models and reality. Rather, "there is a parallelism of structure."[12] Likewise, what Serres traces in Leibniz's notion of a system is "real," it is the idea of such a system that is "realized" by the mathematical models in support of it. What Leibniz refers to with his speculative systematicity (the projected integral of all profiles of a modeled system) is granted a certain *real* darkness, while at the same time the system itself is articulated in brightest clarity by the mathematical models, in an entirely clear and formal manner. Every model hence is a *reduced* model, in the literal sense of the word: a reduction of a totality that is, for experience, always singular and circumstantial, abundantly rich, uncomfortably unsettlingly, and thereby also promising. And yet it must be handled and addressed as being somehow (however mysteriously) "systematic." Faithfulness to real experience and objective thought hence elucidate Serres's decisive question with regard to a universality of culture: "If it is not a miracle, can we build it?"[13]

9

Instead of a Conclusion: The Static Tripod[1]

The Sphinx—What animal stands on four feet at dawn, Oedipus, man who is passing by and who will die if he doesn't reply or find the answer to the riddle? *Oedipus*—Doubtless man, who before walking or standing crawls, a small child, on four legs like an animal. A childish answer. But before man, the animal itself, quadruped like you. Although you lie down in the avenues or before the temples, showing your king's face or your young woman's chest or even spreading your bird's wings, your four legs are obvious to see, oh wildcat. Man and brute mixed can remain quadruped. *The Sphinx*—What animal stands on two feet at noon, beneath the shadowless sun? *Oedipus*—Man, of course, a biped like me, adult, standing, a walker, wandering, with a mobile niche, or like you, with a king's face and queen's breasts, or the animal whose feathered creature's wingspan you display, man therefore and animal too, but this latter flies away, leaving behind he who finally dominates the animals, the intelligent talker, expressing himself because standing straight. *The Sphinx*—What animal stands on three feet when night falls? *Oedipus*—The man, again and always, who leans on a staff of old age when fatigue and age arise. Every animal that walks, to the best of my knowledge, does so on an even number of legs, therefore no beast, no monster, oh Sphinx, could live on three feet. The nonliving, the dead, the inert are necessary for that. Only the object, the thing in equilibrium can stand in front of or after the animal and the man, static tripods, pyramids or tetrahedrons with triangular sides, the results of human labor. They can be called statues, since they stay up all by themselves. Your shadowless questions only bear on statues or equilibria. On the tripod, between us, the incense for the next sacrifice is smoking, and Pythia sometimes comes and sits on it. Three or four feet provide a good seat, not two: man wanders, at Giza, from the Sphinx to the pyramids; these latter will remain, the former will be effaced. But not the staff. The support manufactured by the indefatigable talker, now standing and old, the tool, appeared during the final hours of the formation of this animal who remains a riddle. Oh, Sphinx, did you know that work has three feet?

The Sphinx—Oh, Oedipus, do you know why you're risking death? *Oedipus*—Yes, I've known for a little while now; the decipherers of riddles, my fathers, believed themselves to have gotten out of the difficulty for having heard me answer "man" to your questions. They didn't even consider the fact that we were risking our lives, both of us. If I don't answer or am mistaken you'll kill me; if I say the truth you'll die. We're having a dialog on pain of death. What are we gambling, as though at the dawn of history? Our lives. If I die, you'll sacrifice a man; if you die I'll sacrifice a mixed body of man and animal: here's the first progress. *The Sphinx*—New and unexpected Oedipus among the diviners of riddles of ordinary mothers and fathers, why don't we take up the question again? *Oedipus*—It consists precisely in mixing animal and man. Your riddle resembles your body. It's always necessary to guess the man concealed behind the animal. *The Sphinx*—Give me some time, Oedipus, before my death. *Oedipus*—Forget that man that crawls as a quadruped during his childhood, soon to be standing, senile so quickly. Why not say he's still on four feet when the embalmers lay him out on the alabaster table shaped like a stretched-out lion to empty him of his entrails and organs? What can he be compared to in his mummy wraps? What dull foolishness! *The Sphinx*—Recount again and take your time; save me. *Oedipus*—Here's the time: this day in which the sun rises, like a godsend, running to its zenith and falling to the western horizon, which everyone takes to be a short life, mysteriously measures our entire history and gives the laws of hominization.

The Sphinx—Say the first law. *Oedipus*—The death we risk face to face both of us and which makes us talk or write so long makes us think, drives us to decipher its riddle. Death in general and intraspecific murder: animals know little of them. We find ourselves at risk of death, facing the world and the other, in front of the crowd and before speech. We must give death an answer. *The Sphinx*—Give me an answer. *Oedipus*–Give you an answer. *The Sphinx*—Give you an answer. *Oedipus*—Give me an answer. Here we are before the altar and the mystery, a riddle completely different from the children's guessing game of a moment ago. At this risk and to save its life, humanity at the dawn of time fell upon animals. The great hunts drawn at Lascaux conceal the hunt for man. The latter slowly becomes human by first becoming animal. It transforms into a kind of Sphinx.

The Sphinx—So who am I? *Oedipus*—Crouched all along your wildcat body, you are the first moment of history, when human sacrifice was hesitating before the first law—thou shalt not kill—and when animal sacrifice began to be practiced as a substitute for intraspecific murder. If the sphinxes aren't killed, they'll ravage the land right up to the extinction of men and their group. When Semitic Noah wanted to save himself and his family from the great destruction that would be caused by the waters of the Flood, he built the animal ark so as to hide in their midst, and as a result kept them. Animals must therefore be killed, must therefore be raised or domesticated; wild animals must be eliminated. Hercules labored, a wooden club over his shoulder, and traversed the world, slaughtering birds, lion, hydra, hinds, boar; see him also change into an animal, the lion's fleece on his body and his face hidden at the bottom of its throat, protecting himself under this beginning of clothing. The Egyptians went around nude, above all the women, except for the priests dressed in skins. I recognize you as being a woman beneath that bestial mane, Sphinx, tightly bound or hidden beneath your riddles and appearances. *The Sphinx*—My body, my name. *Oedipus*—Your name, Greek, says at the same time embrace and strangulation, oh monster who brings death but also covering and implication, the condensed, hidden, tightly bound secret. You're named like your paws: talons. *The*

Sphinx—And you're named like your feet. *Oedipus*—Our two names anticipate the riddle. *The Sphinx*—I designate talons, but you know feet; by your knowledge and the words of language, you become man but I remain beast. *Oedipus*—Your body reads like a living hieroglyphic, just as jackal-headed Anubis or ibis-beaked Thoth do, like Heracles beneath his lion skin or Noah hidden in his menagerie. Remember Osiris whose dismembered corpse was scattered, piece by piece, on the Egyptian plain where, at each sacred place, an animal guarded it. And metempsychosis! It's told that the soul migrates into an animal's body according to its merits. Everything became clear from then on, yet everything became reversed, for all at once, men were going to stop sacrificing animals through fear of killing the man bound in them. *The Sphinx*—They had discovered the secret; they had uncovered the hiding place! *Oedipus*—Yes. From that moment on, the delivered man could emerge from his golden animal skins so as to stand upright and naked in the Greek light, in the temples and the public squares, statues on two legs, simply human. The lawful noon rang, the Hellenic zenith of the great abstract discoveries.

The gods were no longer hiding themselves—nor men.

…

The Sphinx—Oh, Oedipus, guess and say a three-footed word. *Oedipus*—The tribunal. The very one before which we're both appearing today, at the article of death. Or the one that we're forming, you, death, and me. *The Sphinx*—Now say or guess a two-footed word. *Oedipus*—The scales or balance, which is what the tribunal amounts to. On both sides of the rock, our two bodies move in disequilibrium on this seesaw. *The Sphinx*—The last word, with one foot? *Oedipus*—The beam—the rod—which is what the balance amounts to, therefore the tribunal; the authority that immediately decides which of the two of us will die. You've only posed riddles of equilibrium, stations or statues, institutions, and now we're reaching this unique needle together, without seat, deprived of statics, unstable, which wanders in space like our two bodies and our two lives, which moves, which doesn't stand, which suddenly falls in the midst of us, like the time of death, the first or final authority. *The Sphinx*—Three, two, one. *Oedipus*—Your life comes to an end at the zero instant.

The Sphinx—The real evening is falling for me: What animal goes on three feet? *Oedipus*—The old man leaning on his cane when age wears him out: The answer to the examination of old presupposed that the day governed by the rhythm of the sun indicated the duration of a life. But that day indexes the sequence of history, as I have said. The aging, experienced generations shape, cut or carve branches or marble, adapt the tool to he who uses or desires it, beat, hit, shoot, hunt, dig, kill, or aid a hesitant gait by means of the new object or even, by digging and beating, decorate. This twilight animal adds together a man and a thing the way the morning animal mixed man with animal. *The Sphinx*—Animals have feet, not hands. In the mixed body, the box is put forward held by human hands. *Oedipus*—Of the three, one foot matters more, leaving the two others to their living parity and to the upright posture that causes hands to be born: the foot that could be said to be orthopedic or false, the prosthesis serving as a support but that can be detached, marking the final and decisive advance of this living being delivered from death by the animal and from the animal by death, risen at noon, soon talking and measuring—at the price of his life—language by things, and bringing death again before the object-box. This living being suddenly recognizes the world. *The Sphinx*—Farewell. *Oedipus*—Stay. Consider, before you, that, and forget, behind you, the old cases. Look, in silence, for a long time at these boxes and these stable pyramids, at this peaceful objective world. What good is it to die, for what archaic causes.

Notes

Preface

1 Michel Serres, *Rome: The First Book of Foundations*, trans. Randolph Burke (London: Bloomsbury, [1983] 2015), 12.

2 Serres, "Vie, information, deuxième principe," in *Serres, La Traduction, Hermes III* (Paris: Les Éditions de Minuit, 1974), 43–72, here 64: "Les tables de nombres remplacent la tragédie. Le hasard n'a plus de projet, il n'a que des combinaisons."

3 Cf. Malcolm Wilson, *Structure and Method in Aristotle's Meteorologica: A More Disorderly Nature* (Cambridge: Cambridge University Press, 2013).

4 Serres, "Vie, information, deuxième principe," in *Serres, La Traduction, Hermes III* (Paris: Les Éditions de Minuit, 1974), 43-72, here 64: "Les tables de nombres remplacent la tragédie. Le hasard n'a plus de projet, il n'a que des combinaisons."

5 Michel Serres, *Pantopie: de Hermes à Petite Poucette. Entretiens avec Martin Legros et Sven Ortoli* (Paris: Le Pommier, 2014), 119.

6 Michel Serres, *L'Hermaphrodite* (Paris: Flammarion, 1987), 96: "Neutre exprime assez bien l'inclusion d'un tiers-exclu: ni l'un ni l'autre ou et l'un et l'autre. La castration joue le rôle d'élément neutre, ici, pour toute opération mettant en jeu l'altérité."

7 We cannot discuss this adequately here. But a point, as will be developed in Chapter 2, is that his concern with invariance and teleonomy (and the primacy of the former to the latter) is exactly a way out of a space that would always already be coordinated. It is why Monod affirms *objective chance—objective chance* is the only way of not semantically importing anything when taking chance into our accounts. Objective chance means no more and no less than countable and measurable chance, chance as an object that can be counted and measured in a great variety of ways.

8 Allen Turing, "On Computable Numbers, with an Application to the Entscheidungsproblem," *Proceedings of the London Mathematical Society* (1937), 42: 230–65; Allan Turing, "The Chemical Basis of Morphogenesis," *Philosophical Transactions of the Royal Society of London* (1952), series B 237, nr. 641: 37–72.

9 Theodore Hailperin, "Boole's Algebra Isn't Boolean Algebra," *Mathematics Magazine* 54, no. 4 (September 1981): 172–84; George Boole, *An Investigation into the Laws of Thought on Which Are Founded the Mathematical Theories of Logic and Probabilities* (Cambridge: Cambridge University Press, [1853] 2009).

10 Michel Serres, Ilya Prigogine and Isabelle Stengers, *Anfänge: Die Dynamik—von Leibniz zu Lukrez* (Berlin: Merve, 1991), xx.

1 Introduction

An editorial note: I have often worked with the original French texts, either because no translations are available, or because I felt the need to provide my own translations. Where I have done this, the French original is provided in the endnotes.

1   Michel Serres, *The Five Senses*, trans. Margaret Sankey and Peter Cowley (Manchester: Continuum, [1998] 2008), 340.

2   Michel Serres, "Structure et importation: des mathématiques aux mythes," in M. Serres (ed.), *Hermes I, La Communication* (Paris: Les Éditions de Minuit, 1968), 21–35, here 34. In the French original: "de comprendre d'un coup le miracle grec des mathématiques et la floraison délirante de leur mythologie."

3   Michel Serres, *Darwin, Bonaparte et le Samaritain, Une philosophie de l'histoire* (Paris: Le Pommier, 2016).

4   Michel Serres, "Structure et importation," 21.

5   Ibid., 34, in the French original: "Notre temps réconcilierait alors la vérité et le sens."

6   Ibid., "cet autre monde dionysiaque."

7   Ibid., "Donner aux figures de cet autre monde dionysiaque des significations épaisses, compactes et obscures où se projettent l'âme humaine, son affectivité, son destin, est juste: il s'agit bien là de la réalité et de la destination de l'homme, de son heur et de ses malheurs, pris universellement."

8   Ibid.

9   Ibid., 34–5: "Mais, outre qu'elles sont des symboles de l'histoire, ne seraient-elles pas dans leur ultime surcharge, dans leur dernière détermination, des modèles signifiants de structures transparentes, de l'ordre de la connaissance, de l'intellect et de la science?"

10 I am citing here from Michel Serres, *Pantopie: de Hermes à Petite Poucette. Entretiens avec Martin Legros et Sven Ortoli* (Paris: Le Pommier, 2014), chapter 8, "Le Grand Fétiche ou les métamorphoses du religieux," 294.

11 For example, France Culture, Masterclasse avec Michel Serres, "Le problème de la violence a été au cœur de ce que j'ai produit depuis 70 livres," August 23, 2017. Available online: https://www.franceculture.fr/emissions/les-masterclasses/michel-serres-le-probleme-de-la-violence-ete-au-coeur-de-ce-que-jai (accessed December 27, 2017).

12 Serres, "Structure et importation," 32, "Cela dit, la notion de structure est une notion formelle. Et voici sa définition, dans laquelle nous insistions sur les thèmes ou l'on fait généralement contresens: une structure est un ensemble opérationnel à signification indéfinie … groupant des éléments, en nombres quelconque dont on ne spécifie pas le contenu, et des relations, en nombre fini, dont on ne spécifie pas la nature, mais dont on définit la fonction et certains résultats quant aux éléments. … Au lieu de symboliser un contenu, un modèle "réalise" une structure. Le terme de structure a cette définition, claire et distincte et pas d´autre."

13 Ibid., 21, "Ce siècle a été le théâtre de plusieurs bouleversements profonds de nos conceptions scientifiques: révolutions accomplies et d'autres à venir, légèrement pressenties, qui font pivoter brusquement l'univers théorique et, avec la lenteur due à leur inertie, le monde de la praxis et les ensembles techniques."

14 Ibid., "une notion méthodique claire, distincte et lumineuse."

15 Ibid., "Nous ne rêvons plus tout à fait les mêmes rêves que nos prédécesseurs immédiats, nous ne pensons ni écrivons comme eux."

16 Michel Serres, *Statues: The Second Book of Foundations*, trans. Randolph Burke (London: Bloomsbury, [1987] 2015), 199.

17 Ibid.

18 Ibid., 48.

19 Ibid., 47.

20 Serres, "Structure et Importation," 27, "Cette leçon ne peut plus être oublie (Bachelard's *Nouvel ésprit scientifique*): historiquement, elle est capitale, car elle ouvre

un nouveau classicisme ou la raison ne tourne plus le dos aux contenus culturels, ou elle ne cherche plus à les comprendre par la médiation d'archétypes symboliques, mais directement, au moyen de ses armes propres, ou elle cherche à mettre en évidence la rigueur structurale de l'amoncellement culturel: c'est pourquoi nous avons dit Logoanalyse."

21 Ibid., 21.
22 Ibid., 47.
23 Michel Serres, "Ce que Thalès a vu au pied des Pyramides," in M. Serres (ed.), *Hermes II, L'Interférence* (Paris: Les Éditions de Minuit, 1972), 163–80, 178–9, "Les formes pures et simples ne sont ni si simples ni si pures, elles ne sont plus des connus théoriques complets, des vus et sus sans résidus, mais des insus théoriques objectifs infiniment répliqués, d'énormes virtualités de noèmes, comme les pierres et les objets du monde, comme nos constructions de pierre et nos objets ouvrés. La forme cache sous sa forme des noyaux de savoir transfinis [178] dont on se prend à craindre que l'histoire ne va pas suffire à les épuiser, des instances fortement inaccessibles comme des tâches. Le réalisme mathématique s'alourdit et reprend une compacité qu'avait dissoute le soleil platonicien. Les idéalités pures ou abstraites font de l'ombre, sont pleines d'ombre, redeviennent noires comme la Pyramide ; la mathématique d'aujourd'hui, quoique maximalement abstraite et pure, se développe dans un lexique issu, pour partie, de la technologie."
24 Serres, *Statues*, 199.
25 Hermann Weyl, "Invariants," *Duke Mathematical Journal* 5, no. 3 (1939): 489–502.
26 Serres, "Ce que Thalès a vu," 179.
27 John Orton, *The Story of Semiconductors* (Oxford: Oxford University Press, 2004), 3.
28 Ibid.
29 Serres, *Pantopie*, 39.
30 Ibid., 46, "J'avais été le témoin, à l'École normale supérieure, du renversement, ou de la transition, si vous préférez, des mathématiques classiques vers le mathématiques modernes" (46), and "J'ai vu, de mes yeux vu, Henri Cartan, qui avait été le grand patron des mathématiques après-guerre, arrive au séminaire Bourbaki, le grand séminaire de mathématique de l'époque, avec Alexandre Grothendieck, véritable génie des mathématiques, à ses côtés. Et j'ai entendu Cartan déclarer qu'à compter de ce jour il serait le secrétaire de Grothendieck, qui était plus jeune d'un quart de siècle." J'ai donc eu la chance de vivre, en temps réel, trois ou quatre grandes révolutions scientifiques: les maths modernes, la biochimie, la théorie de l'information, et plus tard, dans la Silicon Valley, l'arrivée du numérique."
31 Ibid., 39. "Toujours est-il que les autres se sont mis à le chahuter à mort, interrompant la conférence, parce que pour eux, marxistes déterministes, il était l'un des apôtres de l'indéterminisme avec Heisenberg. Il était un ennemi de classe, un réactionnaire à qui il fallait faire la peau! J'ai vu Broglie partir penaud, protégé par deux ou trois personnes, sous les huées des étudiants, Althusser en tête."
32 Ibid., especially chapter 4 entitled "Le Thanatocrate ou le pouvoir de la mort," 161–84; as well as Michel Serres, "Trahison: la thanatocratie," in M. Serres (ed.), *Hermes III, La Traduction* (Paris: Les Éditions de Minuit, 1974), 73–106.
33 Michel Serres, *The Parasite*, trans. L. Schehr and R. Laurence (Baltimore: John Hopkins University Press, [1980] 1982), 5–6.
34 Serres, *Statues*, 51.

35 I discussed my ideas on this in Vera Bühlmann, "Cosmoliteracy and
    Anthropography: An Essay on Michel Serres' Book *The Natural Contract*,"
    in Rick Dolphjin (ed.), *Michel Serres and the Crisis of the Contemporary*
    (London: Bloomsbury, 2018) (forthcoming), as well as in "Foederae Naturae. Eine
    Annäherung zwischen Jean-Luc Nancy's Begriff der Exscription und Michel Serres'
    Begriff der transzendentalen Allgemeinheit von Information" (2015), unpublished
    manuscript. Available online: https://www.academia.edu/31872221/Foederae_
    Naturae._Eine_Annäherung_zwischen_Jean-Luc_Nancy_s_Begriff_der_Exscription_
    und_Michel_Serres_Begriff_der_transzendentalen_Allgemeinheit_von_Information
    (accessed January 7, 2018).
36 Michel Serres, *Le gaucher boiteux: puissance de la pensée* (Paris: Le Pommier,
    2015). Cf. also his recent lecture on this: Michel Serres, "La culture scientifique
    et le numérique." Available online: https://www.youtube.com/watch?v=_
    eyKWOonh1c&t=418s.
37 Michel Serres, *Le tiers-instruit* (Paris: François Bourin, 1991).
38 Michel Serres, *The Natural Contract*, trans. Elizabeth MacArthur and William Paulson
    (Ann Arbor: University of Michigan Press, [1990] 1995).
39 Serres, *Du bonheur, aujourd´hui* (Paris: Le Pommier, 2015), 85. "Je crois aussi que
    la philosophie est á la fois le comble de la culture et le couronnement de l'éducation.
    Elle permet de comprendre la sagesse, la science, etc. Surtout elle permet, à mon avis,
    de prévoir. Et ceci est ma définition propre de la philosophie: c'est l'anticipation des
    pratiques et des théories à venir, des civilisations à venir."
40 Francisco Goya, *Men Fighting with Sticks* (around 1820), in Serres, *Natural Contract*.
41 Serres, *Natural Contract*, 108.
42 Ibid., 92–4.
43 Michel Serres, *Geometry: The Third Book of Foundations*, trans. Randolph Burke
    (London: Bloomsbury, [1993] 2017), 47. It must be noted that the unfortunate title of
    this English translation not only opts for an economical abbreviation of the French
    original (which is entitled *Les Origines de la géométrie* (Paris: Flammarion, 1993)),
    but, in lacking any mention of Serres's pluralization of origins, arguably actively
    misleads the reader.
44 Serres, *Geometry*, 56.
45 Serres, *Natural Contract*, 93.
46 Cf. Serres, *The Five Senses*, here the last episode "Signature," 333–45.
47 Michel Serres, *Rome: The First Book of Foundations* (London: Bloomsbury, [1983]
    2015); *Statues*; *Geometry*.
48 Serres, *Geometry*, 56.
49 Ibid.
50 Ibid.
51 Michel Serres, "Les sciences et la langue," *Rencontres Science et Humanisme*
    (Ajaccio: 2012). Available online: https://www.youtube.com/watch?v=JV8E5w3nClk
    (accessed December 27, 2017).
52 Ibid., 9, "*De senectute vitæ:* here then is a true oldness, common to the dying and
    the newborn, to little girls and grandmothers, to animals and plants, to friends, to
    enemies, all of them bearers of a DNA, all of them equal in time, with the exception of
    two parts, the one minimal, their individual age, the other much bigger, the interval
    lived since their species made its appearance. I'm not taking into account thought,
    emotions or cultures, narrow and lightning-fast."

2 Quantum Literacy

1 Serres, *Rome*, 10.
2 Gilbert Simondon, "Save the Technical Object," interview by Anita Kechickian, an English translation by Andrew Illiadis of a 1983 interview that Simondon gave to the French magazine *Esprit* (*Esprit*, April 1983, 147–52).
3 Ibid.
4 Jean-François Lyotard, "Le non sens commun," cited from the English translation: "Domus and the Megalopolis," in J.-F. Lyotard (ed.), trans. Geoffrey Bennington and Rachel Bowlby, *The Inhuman: Reflections on Time* (London: Polity Press, 1991), 204.
5 There is, of course, a certain mysticism to this, and Serres readily acknowledges that this is the main lesson he learnt from Simone Weil, cf. Serres, *Pantopie*, 34. Serres also calls himself "a mystic of mathematics," in Serres, "Yeux," talk at the Institute Catholique de Paris, December 18, 2014. Available online: https://www.youtube.com/watch?v=1qFdYgjWg9s (accessed January 5, 2018).
6 Ibid.
7 The book where Michel Serres develops this most is *Le gaucher boiteux: puissance de la pensée* (Paris: Le Pommier, 2015).
8 This notion of "import" is central to Chapter 7 in the present volume, entitled "Sophistication and Anamnesis: Remembering an Abundant Past."
9 Cf. with regard to the idea of universal genitality as the spiritual neutrality of hermaphrodite Michel Serres, *L'Hermaphrodite* (Paris: Flammarion, 1987), 72, "Start from antithesis, and you will reach castration" he maintains in his book on Balzac's story "Sarrasine." "Start from enantionmorphy (chirality) of right and left ... start from symmetry, and you will produce the Hermaphrodite," 72. This same theme, a notion of reason that operates in terms of chirality, is also central to Serres, *Le gaucher boiteux*.
10 Michel Serres, "Interview de Michel Serres sur l'autorité," APEL, the association of parents of school kids in France, in the context of a conference titled "Autoriser l'autorité," in Montpellier 2010. Available online: https://www.youtube.com/watch?v=GeJgTmZ9EBc (accessed January 8, 2018).
11 Cf. also Serres, *L'Hermaphrodite*, especially 70, as well as *Le gaucher boiteux*.
12 "I think I can safely say that nobody understands quantum mechanics," in Richard Feynman, *The Character of Physical Law* (Cambridge: MIT Press, [1964] 2017), 129.
13 Feynman, *Character of Physical Law*, 129.
14 Ibid.
15 Ibid.
16 Ibid.
17 Ibid.
18 Ibid.
19 Ibid.
20 Ibid.
21 Richard Feynman, *QED: The Strange Theory of Light and Matter* (Princeton: Princeton University Press, [1986] 2014).
22 Feynman, *Character of Physical Law*, 129.
23 Ibid.
24 Feynman, *QED*.
25 Serres, *Rome*, part one ("The Black and White: The Covering Over").

26 Cf. especially: Michel Serres, *The Incandescent*; *The Natural Contract*; *Rameux* (Paris: Le Pommier, 2004). The term "Natural Communication" is currently coined by the mathematician and theoretical physicist Elias Zafiris in his forthcoming book *Natural Communication: A Functorial Approach*; Zafiris's approach has been the basis for Nikola Marincic's study on Louis Hijelmslev's glossematics in relation to thinking about coding as literacy: *Towards Communication in CAAD. Spectral Characterisation and Modelling with Conjugate Symbolic Domains* (Zurich: ETH Zurich PHD Thesis, 2017).

27 Serres, *Rome*, 10.

28 Ibid.

29 Ibid.

30 Michel Serres, *La Légende des Anges* (Paris: Flammarion, 1993).

31 Cf. Serres, *Pantopie*, 119.

32 Ibid., 152, "Quand l'épistémologie réflexif devient intrinsèque, le champ transcendantal passe à l'objectif." There is a superb introduction to Serres transcendental philosophy by Anne Crahay, *Michel Serres. La mutation du cogito. Genèse du transcendental objectif* (Bruxelles: de Boeck, 1988).

33 As also Baruch de Spinoza saw it in his *Ethics* (1677).

34 Cf. The last chapter, "The Statue of Hestia," in *Statues*, which Serres dedicates to epistemology (198–201).

35 Serres, *Pantopie*, 152: "Elles passent dans le monde. Ce n'est plus nous qui informons les choses avec nos schèmes et nos catégories subjective, le monde est information."

36 Ibid., "La nouvelle science des chose fait qu'on est bien obligé de faire cette philosophie-là."

37 Ibid., "C'est parce-que précisément, la chlorophylle a telle et telle formule chimique, qui est de l'ordre de l'information, qu'elle peut transformer la lumière en matière."

38 We will elaborate on such an understanding of code-relative description later, as that of alphanumerical code, in Chapter 5, under the subheading "Quantum Writing: Substitutes Step in to Address Things Themselves." There is also an interesting kinship between such code-relative understanding of "description" and what Jean-Luc Nancy has been thinking about as "exscription"; cf. Bühlmann, "Foederae Naturae."

39 Michel Serres, *Genesis* (Ann Arbor: University of Michigan Press, [1982] 1995), 9.

40 Ibid. Serres recounts how he originally wanted to call this book, for this very reason (to emphasize the fecundity of pure multiplicitousness in its full ambiguity) *Noise* rather than *Genesis* (Serres, *Pantopie*, 183). Noise, in information theory, plays a role that is generative, supportive, as well as corruptive for an entropic notion of order.

41 "Ichnography contains the possible" (21), "ichnography has every direction" (20), "And now, decipher that ichnography" (20) in Serres, *Rome*. In *Genesis*, Serres develops this idea in a dedicated chapter: "Often we are drowned in this confused minuteness. The more one ascends, on the contrary, the flights of integration, the more the rational rationalizes itself. Just as our body integrates the noise of minute perceptions into sensible signals, so does God integrate in absolute knowledge, in white light, the relative noise of our right, flighty thinking. Harmony removes itself from noise, irenism removes itself from fury, as the universal removes itself from the local, the same distance: huge, infinite, measurable. Ichnography, then, should be pure. Smooth, white, united like a perfect chord" (19–20).

42 Serres, *Genesis*, 19–21, the ichnographic painting that captures the act (the nude) only by depicting the index of an indefinite trace (the footprint) in Frenhofer's masterpiece.

Ichnography, the ground plan in architectonic models, turns chromatic in Serres's architectonic of the objective world. Cf. Vera Bühlmann, " 'Ichnography'—The Nude and Its Model. The Alphabetic Absolute and Storytelling in the Grammatical Case of the Cryptographic Locative," in Vera Bühlmann et al. (eds.), *Coding as Literacy, Metalithikum IV* (Basel/Vienna: Birkhäuser, 2015), 276–350.

43 Michel Serres, "Anamnèses mathématiques," in *Hermes I. La communication* (Paris: Les Editions de Minuit, 1968), 78–112, here 82, "Bref, l'élément atomique formel traductible partout dans le système est aussi un condensé d'histoire, enveloppant son origine radicale, la loi de sa série évolutive, et l'horizon de sa finalité."

44 As, for example, Jacques Monod suggests, one of the crucial and explicitly acknowledged influences on Serres, in *Chance and Necessity: An Essay on the Natural Philosophy of Modern Biology* (New York: Vintage, [1970] 1972), 100–1.

45 Michel Serres, *The Birth of Physics*, trans. David Webb and William Ross (London: Rowman and Littlefield, [1977] 2018), kindle edition, position 3209.

46 Serres, *Birth of Physics*, position 3182.

47 Ibid.

48 Ibid.

49 Ibid.

50 For an extensive discussion of mathematics' abstractness, cf. Alfred North Whitehead, *Introduction to Mathematics* (New York: Henry Holt, 1911), especially the first chapter "The Abstract Nature of Mathematics." How this role can be played today by a mathematical notion of information remains a largely open issue to this day.

51 Alfred North Whitehead, *A Treatise on Universal Algebra* (Cambridge: Cambridge University Press, 1998).

52 Ibid., vi.

53 Ibid. Cf. also the discussion of how Universal Algebra proceeded and evolved until the 1960s by George Grätzer, "Universal Algebra," in M. H. Stone, L. Nirenberg and S. S. Chern (eds.), *University Series in Higher Mathematics* (Princeton: D. van Nostrand, 1968); as well as, for a discussion of the subject's developments since the 1950s until 2012, Fernando Zalamea *Synthetic Philosophy of Contemporary Mathematics* (London: Urbanomic. 2012), and the critical appreciation of Zalamea's book by Giuseppe Longo, "Synthetic Philosophy of Mathematics and Natural Sciences" (CNRS, Collège de France et École Normale Supérieure, Paris, 2015). Available online: https://www.di.ens.fr/users/longo/files/PhilosophyAndCognition/Review-Zalamea-Grothendieck.pdf (accessed February 28, 2016).

54 It is for precisely this reason that Lyotard has situated the stakes for the condition of knowledge in computerized societies as the central problem that triggers a break with the modern traditions: Jean-François Lyotard, *The Postmodern Condition: A Report on Knowledge* (Minneapolis: University of Minnesota Press, [1979] 1984).

55 Cf. Sören Stenlund, *The Origin of Symbolic Mathematics and the End of the Science of Quantity* (Uppsala: Uppsala University Press, 2014).

56 Cf. ibid.

57 Indeed, the separation of a particular notational system for mathematics is a rather recent development compared to the history of mathematics; it is a bifurcation after many centuries of using one and the same script for linguistic as well as mathematical articulations. Cf. Gerd Schubring, "From Pebbles to Digital Signs: The Joint Origin of Signs for Numbers and for Scripture, Their Intercultural Standardization and Their Renewed Conjunction in the Digital Era," in Vera Bühlmann and Ludger Hovestadt (eds.), *Symbolizing Existence, Metalithikum III* (Vienna: Birkhäuser, 2016).

58 Cf. Rosa Massa Esteve, "Symbolic Language in Early Modern Mathematics: The Algebra of Pierre Hérigone (1580–1643)," in *Historia Mathematica*, 35, no. 4, November 2008, 285–301. Serres points out the relevance of this tradition for his thinking when he writes in *The Incandescent*, 59–60. "At the origin of algebra, a thing (the *cosa*, as the first Italian algebraists said) called x takes on every value because it doesn't have any value of its own. Again a symbol, again a white token or a general equivalent. We already carry in our incandescence the potential origin of our knowledge; we have already said of the human: this unknown = x."

59 Cf. www.etymonline.com; also Wilhelm G. Jacobs, "Das Absolute," in Hans Jörg Sandkühler (ed.), *Enzyklopädie Philosophie* (Hamburg: Felix Meiner Verlag, 2010), 13–17.

60 It seems not entirely implausible, at least, to think about the early twentieth century's foundational crisis as a continuation of these very disputes on higher levels of abstraction. For a largely unbiased account and a serious and open-minded suggestion of how to approach the dilemma, cf. Hermann Weyl, *The Continuum: A Critical Examination of the Foundation of Analysis* (London: Dover ([1918] 1994).

61 Due to its brevity, this summary follows the lineage in the Kantian tradition, which has turned out to be the predominant one throughout the twentieth century. There also is a more empiricist tradition within the history of algebra. The account here inevitably understates the ideas of Leibniz, Lambert, and others, who maintained that algebra, by its calculatory and symbolic methods that cannot be reduced to arithmetic, could actually be seen as opening up the closedness of classical logics in the Aristotelian tradition, thereby introducing an *ars inveniendi*, an art of invention into logics—an approach that was still pursued by figures as eminent for the nineteenth-century development of algebra as Charles Sanders Peirce (abduction) or Richard Dedekind (abstraction as a creative act). Against these ideas, Kant famously stated, "Die Logik ist … keine allgemeine Erfindungskunst und kein Organon der Wahrheit; keine Algebra, mit deren Hülfe sich verborgene Wahrheiten entdecken ließen"(Immanuel Kant, *Logik. Ein Handbuch zu Vorlesungen* (1800), Königsberg: Friedrich Nicolovius; Akademie-Ausgabe, Bd. 9, 1–150, A 17, here cited in Volker Peckhaus, "Die Aktualität der Logik als Organon," in G. Abel (ed.), *Kreativität: XX. Deutscher Kongress für Philosophie* (September 26–30, 2005), an der Technischen Universität Berlin: Kolloquienbeiträge (Hamburg: Felix Meiner Verlag, 2006). For an introduction to the conflicts the unsettled status of algebra triggered in the empirical sciences themselves, cf. Isabelle Stengers, *Cosmopolitics I* (Minneapolis: University of Minnesota Press, [2003] 2010), especially book 2, *The Invention of Mechanics, Power and Reason*, therein "The Lagrangian Event" (112–28) and "Abstract Measurement: Putting Things to Work" (129–38).

62 Cf. the seminal study on the implications of this for axiomatics by Robert Blanché, *L'Axiomatique* (Paris: Presses Universitaires de France, [1955] 1980); there is an English edition of this book by Geoffrey Keene (New York: Free Press of Glencoe, 1962), but it excludes the crucial two chapters with which Blanché ends his study, discussing the implications for science and for philosophy. Cf. on the genealogy of philosophical notions of necessity and contingency, and the relatively recent upheaval with regard to this genealogy, also Jules Vuillemin, *La Philosophie de l'Algèbre, Tome I: Recherches sur quelques concepts et méthodes de l'Algèbre modern* (Paris: Presse Universitaire, 1996 and [1962] 2006); Jules Vuillemin, *Necessity or Contingency. The Master Argument* (Chicago: University of Chicago Press, 1996). Jules Vuillemin, *What Are Philosophical Systems?* (Cambridge: Cambridge University Press, 2009).

63 Serres, *Birth of Physics*, position 3202.

64 Ibid.

65 Michel Serres, "Vie, information, deuxième principe," in Serres (ed.), *La Traduction, Hermes III* (Paris: Les Éditions de Minuit, 1974), 43–72, here 54: "pour ne pas risquer de confondre invariance et identité."

66 Algebra, which used to be regarded in the seventeenth and eighteenth centuries as the theory of equations, was transformed by the early twentieth century into what was now called Quantics, the theory of "algebraic forms," also called "residual forms." Residual forms are forms that "define" by conserving something indefinite throughout transformations, while it is the transformations themselves that regulate their formality, their morphisms (not the nature of any sort of content, as in the hylemorphic tradition). With the seminal work of Emmy Noether and others on reformulating the laws of thermodynamics as laws of conservation, Quantics turned into a general theory of invariances. Cf. Yvette Kosmann-Schwarzback, *The Noether Theorems. Invariance and Conservation Laws in the Twentieth Century*, trans. Bertram E. Schwarzbach (New York: Springer, 2011); Jean Martin Levy-Leblond and Françoise Balibar, *Quantics: Rudiments of Quantum Physics* (Amsterdam: North Holland, 1990).

67 Monod, *Chance and Necessity*, 101.

68 The author [Monod] here refers his reader to V. Weisskopf, in "Symmetry and Function in Biological Systems at the Macromolecular Level," in Arne Engstrom and Bror Strandberg (eds.), *Nobel Symposium No. 11* (New York, 1969), 28.

69 Monod, *Chance and Necessity*, 101.

70 Serres, "Vie, information, deuxième principe," 63, "Il n'est pas de bonne méthode, pour les hommes de science, de prendre aux philosophes une définition du hasard. Car il est Dieu ou rien, selon que le penseur se tient ou pour rien ou pour Dieu. Or, nul ne peut rien savoir sur rien avant et sans la science lorsqu'elle a déjà pensé quelque chose; aux philosophes, donc, de chercher la définition auprès d'elle."

71 Ibid., 63, "Avant le dix-septième siècle, on vivait le destin et la fatalité, stoïque, mahométane, chrétienne, on subissait les heurs et les revers, on craignait la fortune et le sort, on s'émerveillait des rencontres, des occasions et des traverses, on pratiquait les jeux, dés, oie ou solitaire, on invoquait les dieux, soumis eux-mêmes à des pouvoirs aveugles, on mourait du vent, du naufrage; le monde était intentionnellement tissé, inattendu, cruel et nécessaire: on ne connaissait pas le hasard."

72 Ibid., "Comme objet, c'est-à-dire vidé à tout jamais de quelque trace subjective."

73 Ibid.

74 Serres, "Vie, information, deuxième principe," 63, "Les tables de nombres remplacent la tragédie. Le hasard n'a plus de projet, il n'a que des combinaisons. C'est, si l'on veut, le postulat d'objectivité."

75 Ibid., "Depuis lors, ce hasard bien formé déserta passablement les philosophies: elles le supportaient aussi mal que les religions autrefois, et cela est un signe."

76 Ibid., 64, "Le monde est-il saturé de liaisons ou localement lacunaire, c'est une question à laquelle nous avons ou trop ou trop peu de réponses."

77 Ibid., "Quelque chose comme une antinomie, au sens kantien, pour le moment indécidable. La définition est moins claire que le défini: celui-ci permet un calcul et celle-là des hypotheses."

78 Ibid., 65, "que le hasard, dans le pluriel des cas, s'énonce d'un mélange cent fois plus que d'indépendances, que d'indépendances individuées."

79 Serres uses *ensemble* in French, which he specifies as "une variété de base." This is why I do not translate *ensemble* here as "set"—a set is not, in the mathematical sense, a

variety, that is, a collection of solutions to a system of polynomial equations (over the real or the complex numbers). A "set" in the sense of set theory is simply a collection of distinct objects.

80 Serres, "Vie, information, deuxième principe," 65, "D'où ce premier résultat: le hasard serait impensable—seulement pathétique—sans un ensemble, une variété de base. L'important, ici, est bien le pluriel, le nombre des 'facteurs', la pure multiplicité. A l'ancien monde, toujours déjà légalisé, fibré de séquences toutes prêtes, se substitue un tableau, une collection chiffrée quelconque, un paquet de termes sur lesquels, justement, on ne fait aucune hypothèse. La chose est vérifiable en tous lieux. En mathématiques, cet ensemble fut nommé collectif."

81 Ibid., 66, "L'événement fortuit, quel qu'il soit, est figure sur fond, sur collectif de fond: et ce fond n'est pas un cosmos, c'est un nuage; qu'il soit immense, il n'est plus dominé: le chaos."

82 Cf. Serres, *Genesis*.

83 Serres, "Vie, information, deuxième principe," 65.

84 Serres, "Vie, information, deuxieme principe."

85 David L. Watson in his 1930 article entitled "Entropy and Organization," *Science*, (August 29, 1930): 220–2, cited in James Gleick, *Information: A History, a Theory, a Flood* (London: Harper Collins, 2011), here from the Kindle edition: position 4306.

86 Erwin Schrödinger, *What Is Life? The Physical Aspect of the Living Cell* (Cambridge: Cambridge University Press, 1944).

87 Claude Shannon, "A Mathematical Theory of Communication," *Bell System Technical Journal* 27, no. 3 (1948): 379–423.

88 Léon Brillouin, *Science and Information Theory* (New York: Academic Press, [1956] 2013).

89 David L. Watson in his 1930 article entitled "Entropy and Organization," *Science*, (August 29, 1930): 220–2, cited in James Gleick, *Information: A History, a Theory, a Flood* (London: Harper Collins, 2011), here from the Kindle edition: position 4306.

90 Gleick, *Information*, Kindle edition: position 4313.

91 Ibid., position 4323.

92 Ibid.

93 Brillouin, cited in C. Gaither and A. Cavazos-Gaither (eds.), *Gaither's Dictionary of Scientific Quotations* (New York: Springer, 2012), 1626.

94 Cf. Brillouin, *Science and Information Theory*, Kindle edition: position 2766.

95 Gleick, *Information*, Kindle edition: position 4355.

96 Ibid.

97 In other words, as Serres asks, "le deuxième principe, lui-même, est-il universel? Oui, mais pas tout à fait à la manière de Newton. Il l'est, si j'ose dire, de façon non continue, de place en place. Il y a des archipels: ici et là, parmi eux, des îlots de néguentropie. A la limite, c'est une antinomie, au sens kantien, que de poser ouvert ou fermé l'univers comme tel, du moins pour le moment. En tout cas, il est universel dans sa négation, ou, mieux, dans ce qu'il interdit: le mouvement perpétuel." In my own translation: "Is the second law universal? Yes, but not completely as Newton would have it. It [the second law] is [universal], if I may say so, non-continuously, from region to region. There are archipelagos: here and there between them, islands of negentropy. At its limits, it's an antinomy, in the Kantian sense, to consider a universe open or closed in such a fashion. Either way, it is universal in its negation, or, better, in that which it prohibits: perpetual movement" (Serres, "Vie, information, le deuxième principe," 61).

98　Brillouin, *Science and Information Theory*: "There is no continuity at the atomic level but only discrete stable (or metastable) structures, and the atomic system suddenly jumps form one structure to another one, while absorbing or emitting energy. Each of these discrete configurations of the quantized physical system was called a 'complexion' by Planck." Kindle edition: position 2762.

99　Allan M. Turing, "The Chemical Basis of Morphogenesis" Philosophical Transactions of the Royal Society of London, series B 237, nr. 641: 37–72 and "On Computable Numbers, with an Application to the Entscheidungsproblem," Proceedings of the London Mathematical Society (1937), 42: 230–65.

100　Shannon, "A Mathematical Theory of Communication."

101　Norbert Wiener, *Cybernetics: Or Control and Communication in the Animal and the Machine* (Cambridge: MIT Press, 1948).

102　Shannon discusses the term "negative entropy," but considers its distinction negligible for information as a mathematical quantity notion. It was Norbert Wiener, who via the work by John von Neumann, Alan Turing, Claude Shannon and Leo Szilard maintained against Shannon that negentropy is in fact crucial rather than negligible for a mathematical theory of information. It is largely due to this dispute that until today, different notions of mathematical information are in usage: (1) information as a measure for order in terms of entropy; and (2) information as a measure for order as negentropy. While both speak of information as a measure, and hence capable of establishing order, the two concepts of order are actually inverse to each other: order as negentropy means minimal entropy (maximal amount of bound energy, minimal of free or available energy in Schrödinger's terms), while order as entropy means minimal negentropy (maximal amount of free and available energy, minimal amount of bound energy in Schrödinger's terms). Much confusion in the understanding of "information" arises from this still today. Cf. Gleick, *Information*, Kindle edition: around position 3956. Although, it must be argued, Gleick does not seem to be aware fully of the implications of the issue at stake.

103　Dennis Gabor, MIT Lectures, 1951, cited in Brillouin, *Science and Information Theory*, Kindle edition: position 3805.

104　Brillouin, *Science and Information Theory*, Kindle edition: position 3805.

105　This is something Serres develops further, for example, in *The Incandescent*, here chapter "Descent into Incandescence," 53–4.

106　Cf. regarding such scalar *niveaux* also the writings of the astro-chemist Hubert Reeves, for example, *Origins. Cosmos, Earth and Mankind* (New York: Arcade, [1996] 2011).

107　Cf. The first chapter in Monod's book *Chance and Necessity*, entitled "Of Animisms and Spiritualisms," 23–44, where Monod discusses extensively the distinction between animalism, spiritualism and the role of objective chance in probabilistic reasoning plays. The affirmation of chance as anything other than an effect of limited subjective faculties can of course be called a form of spirituality too. I will even suggest to speak of the incandescent Paraclete with regard to it. But what distinguishes chance as *objective,* and hence such spirituality from any form of spiritualism, is that it depends upon no hypothesis in particular: if it must be called spiritual, then it should be done in the way discussed above which decouples the spiritual from the sacred (sacrifice). Indeed, the affirmation of objective chance is the very idea that allows such a decoupling to be maintained reasonably.

108　This is an idea in development by Elias Zafiris, in his Natural Communication paradigm: Zafiris, *Natural Communication: A Functorial Approach* (Basel/

Vienne: Birkhäuser, forthcoming). Cf. also: E. Zafiris, "The Nature of Local/Global Distinctions, Group Actions, and Phases: A Sheaf-Theoretic Approach to Quantum Geometric Spectra," in Bühlmann et al. (eds.), *Coding as Literacy, Metalithikum IV* (Vienna: Birkhäuser, 2015), 172–86.

109   Rather than trying to distinguish artificial things from things natural, it is possible to quantify "the strangeness of an object," as Monod proposes in *Chance and Necessity*, especially the introduction to the book entitled "Of Strange Objects," 3–22.

110   As my theoretical physicist friend Elias Zafiris explained to me in private conversation. The new number refers to the increased scale on which nuclear science empirically observes particle behavior. In the case of 10^{32} it is the coefficient allowing the famous Higgs Boson to be traced in the CERN Accelerator.

111   Architects in earlier days, who were responsible for the building of cathedrals, had to commit themselves literally with their own life: delaying the progress of the building through accident or mistake was to be paid for by the architect's life.

112   Serres, *Statues*, 5.

113   Serres, "Vie, information, deuxième principe," 54, "pour ne pas risquer de confondre invariance et identité."

114   Cf. Serres, *Pantopie*, 52.

115   Serres, "Vie, information, deuxième principe," 49.

116   Cf. the introductory chapter to Monod's *Chance and Necessity*, entitled "Strange Objects," 3–22.

117   Serres, "Vie, information, deuxième principe," 47, "le génomène est le secret codé du phénomène."

118   Monod, *Chance and Necessity*, 114 (emphasis in the original).

119   Serres, *Birth of Physics*, position 3202. Serres further developed this idea of a ciphered atomism in his more recent book *The Incandescent* (Serres, [2003] 2018).

120   Cf. Serres, "Vie, information, deuxième principe," 46.

121   Monod, *Chance and Necessity*, 14.

122   This is his commitment to Brillouin's emphasis on the substantiality of code in information theory, which led him to distinguish negative and positive information alongside the distinction between negative and positive entropy. Cf. Brillouin, *Science and Information Theory*, and Serres's discussion of this commitment at length in his article "Vie, information, deuxième principe."

123   Michel Serres, *Le Système de Leibniz et ses modèles mathématiques* (Paris: Presses universitaires de France, 1968).

124   Serres, *Birth of Physics*, position 3142–82.

125   Ibid., 63, "Est-il si surprenant qu'il y aille de la pure multiplicité? Le hasard n'est pas quelque chose et il n'a nulle dimension, tout comme la chaleur et l'information. Pensez-le comme une chose et donnez-lui des dimensions, vous trouverez un monde lié ou un dieu au-dessus des dieux, savoir la nécessité, son contraire: une anti-physique, une métaphysique. Non, le hasard est nombre, jeu de nombres. Il est même écrit dans les nombres."

126   Ibid., 46, "les phénomènes les plus improbables, lorsqu'ils existent, se reproduisent; ils se produisent parce qu'ils se reproduisent."

127   Ibid., 49, "Le physicien est devenu anarchiste, le bruit de fond est son domaine, et la musique est rare. Tout objet se nomme miracle. Le vieil étonnement a renversé les rôles. Il faut accepter ce retournement, auprès de quoi le copernicien est un jeu d'enfants, pour entrer en science, aujourd'hui. Qu'il soit insupportable à nos stéréotypes culturels et au goût que nous prenons d'habiter notre propre maison,

c'est l'évidence. Et pourtant, elle tourne. Cela se traduit aujourd'hui: et pourtant, il est extraordinaire qu'elle tourne. La loi n'est plus dans la loi ni l'homme dans le monde. Il est incompréhensible qu'il y ait du compréhensible. Et pourtant, donc, il y a de l'ordre. Des conservatoires de hasard et des systèmes auto-réglées, bourrés jusqu'à la gueule de néguentropie."

128 Berhard Stiegler has recently mentioned in one of his talks a very important point with regard to the role of such acceptance. It took Albert Einstein several years, he said, until he himself could accept the idea of general relativity.

129 Serres, *Leibniz*, 4, "Le système est optimiste, pourrons nous en reconstruire un aussi beau, parmi les bruits et la fureur?"

130 Serres, *Pantopie*, 225–6: "Oui, matériel. Avec une couleur, avec du feu, et avec la référence concrète d'une personnage concret, vivant: L'Incandescent ou l'Hominescent."

131 I have developed this idea of how to think about spectral characters in terms of what in information science is called an electrotechnical channel in Vera Bühlmann, "Generic Mediality: On the Role of Ciphers and Vicarious Symbols in an Extended Sense of Code-based 'Alphabeticity,'" in Rosi Braidotti and Rick Dolphijn, *Philosophy after Nature* (London: Rowman and Littlefield, 2017), 31–54. For a discussion by Serres of his method of "inventer des personnage," cf. Serres, *Pantopie*, chapter 2 entitled "Pantope ou penser en inventant des personnages," 69–114. It would be interesting to compare Serres's notion, with regard to the spectral character of his persona, with the notion of Conceptual Persona in Deleuze and Guattari. See, for example, Gregg Lambert, *Philosophy after Friendship. Deleuze's Conceptual Personae* (Minneapolis: University of Minnesota Press, 2017).

132 Serres, *Pantopie*, 226, "Il [l' Univers] est escent. Il est inchoatif. Il n'est pas blanc. Il est de toutes les couleurs. Il est plutôt noir que blanc, même. Mais il est brûlant, ça il n y a pas de doute. Il vient d'un feu primitive. Ce qu'il y a d'intéressant dans le concept d'incandescence, c'est justement qu'il peut s'appliquer aussi bien au big bang, il y a quinze milliards d'année, qu'à un événement atomique ou subatomique qui va se dérouler sur 10^{-15} secondes. C'est vraiment un concept qui permet de toucher toutes les dimensions du temps."

133 Michel Serres, "L'interférence théorique: tabulation et complexité," in *Hermes II, L'Interférence* (Paris: Éditions de Minuit, 1972), 33, "Il n'y a plus de sphinx interrogeant, au détour de la route, à qui je dois, seul, arracher son savoir, par la patience et la mobilité de mes suppositions, par la pertinence et l'astuce de mes hypothèses."

134 Ibid., "l'intersubjectivité à ce démon subtil qui n'est pas tricheur, et qui est le monde même: il n'est pas devant moi, il nous enveloppe et nous l'investissons, nous y sommes plongés ou il est cerné par le graphe complexe du labyrinthe encyclopédique."

135 Serres, *Statues*.

136 Serres, *Birth of Physics*, position 3018 (emphasis in the original).

137 Michel Serres, "La querelle des Anciens et des Modernes," in *Hermes I, La Communication* (Paris: Éditions de Minuit, 1968), 46–77. "Elle [l'épistémologie] perd lentement le champ original de ses problèmes au bénéfice de la technique scientifique qui, de son côté, commence à les prendre en charge," 77.

138 Serres, "La querelle des Anciens et des Modernes," 77, "l'épistémologie alors s'historicise: elle devient régionale (elle explose et se distribue en descriptions partielle de champs de plus en plus étroits), elle devient impressionniste (elle décrit

de plus en plus précisément la région en question, rejetant toujours à plus tard l'entreprise gnoséologique), bref, elle se constitue en histoire naturelle des sciences."

139 Ibid.
140 Ibid., 75, "Les temps modernes survienne dès que les deux fils s'infléchissent sensiblement l'un vers l'autre. Ce concours a, nous le savons, deux raisons: l'épistémologie échoue sur le terrain de ses anciennes victoires, la mathématique prend le goût de triompher sur le champ de ses récentes défaites. … Les problèmes sont les mêmes, mais technicisés, mais formalisés, mais épurés de leur aura réflexive."
141 Ibid., 76, "Aucun problème ne peut jamais être considéré comme définitivement résolu; il y a un 'historicisme' essentiel qui fait que la mathématique est un mouvement autant qu'un système."
142 Serres, *Natural Contract*, last chapter, "Casting Off," 97–124.
143 Serres, *The Five Senses*, 340.
144 Michel Serres, 'Les sciences et la langue,' Rencontres Science et Humanisme 2012, Ajaccio. Available online: https://www.youtube.com/watch?v=JV8E5w3nClk (accessed December 27, 2017).
145 Serres, *Statues*, 22.
146 Ibid., 22–3.
147 Serres, *The Incandescent*, 54.
148 In German, I would translate the term "puissance" as *Mächtigkeit*.
149 Serres, "Les sciences et la langue."
150 Ibid.
151 Cf. Immanuel Kant, "Amphiboly of Concepts of Reflection," appendix to "The Transcendental Analytic" of *Critique of Pure Reason*. Cf. the essay on Kant's own approach to what he called "Transcendental Deliberation" by Andrew Brook and Jennifer McRobert, "Kant's Attack on the Amphiboly of the Concepts of Reflection," *Theory of Knowledge*, August 1998. Available online: https://www.bu.edu/wcp/Papers/TKno/TKnoBroo.htm (accessed January 12, 2018).
152 Serres, "Les sciences et la langue."

3 Chronopedia I: Counting Time

1 Michel Serres, *Atlas* (Paris: Flammarion, [1994] 1996), 98; "Le temps qu'il fait ou qu'il va faire somme donc celui qui court de la cause vers l'effet, plus celui des probabilités, enfin d'autres que l'on pourrait distinguer en éventail ou bifurquant, linéaire et circulaire; il cumule donc ceux de Newton, de Bolzmann, de Bergson—déterministe, entropique et statistique ou porteur d'improbables nouveautés—. Plus, peut etre, que celui du chaos. Le temps qu'il fait, somme-t-il toutes sort de temps mesurable?".
2 Serres, *Geometry*, xxxiii (emphasis in the original).
3 Ibid., xxxiv.
4 Ibid.
5 Ibid.
6 Ibid. (emphasis in the original).
7 Ibid., xxxv.
8 Ibid., xxxvi (Emphasis in the original).
9 Ibid.

10 Serres, *Statues*, 51, "Let's introduce mass into philosophy, in a way that's compatible with physics and the other sciences and placed by them among the fundamental units in dimensional equations, at the same rank as space and time: all three units counted by them as pure quantities. They measure dimensions without inquiring into their nature."

11 Serres, *Geometry*, xxxv, "Signifying physically and at the origin to percolate, the verb 'to flow' reduces, in the simple and laminar flux, to a particular case. What we took to be the common and reasonable current amounts to a rarity."

12 Cf., for example, Harry Kesten, *Percolation Theory for Mathematicians* (Boston: Springer, 1982).

13 Serres, *Geometry*, xxxvi.

14 Ibid.

15 Ibid.

16 Michel Serres, *L' Incandescent* (Paris: Le Pommier, 2003), 14–15: "Ouvrez donc les yeux nouvellement. Vous voyez moins de l'espace que du temps. Vous voyez moins d'objets disposés dans une étendue familière, fleuves, roches, sommets ou soleil, que différents rythmes d'un écoulement, travaux éphémères, maisons séculaires, rives millénaires, rochers millionnaires, astres milliardaires. Alors que la représentation commune faisait disparaitre le temps dans l'espace, le dissolvait ou mieux encore le dissimulait comme un magicien cache un vol de colombes sous un voile blanc, alors que la scène théâtrale de la représentation rendait difficile à saint Augustin, à Bergson, comme à leurs successeurs, la vision directe, l'intuition ou la pensée de la durée, cette série qui, maintenant, se déplie en faisant jaillir devant moi des millions de fontaines, foudroyantes ou d'une infinie lenteur, devant la maison de compagne ou le Grand Canyon, fait disparaître, dissout, dissimule, à son tour, l'étendue derrière les surgissements d'autant de rythmes chroniques. … Le temps se perdait dans l'espace, voici que l'espace, à son tour, s'engloutit dans le temps."

17 Cf., for example, Hans-Dieter Bahr, *Über den Umgang mit Maschinen* (Tübingen: Konkursbuch, 1983).

18 https://www.etymonline.com/word/scale.

19 Serres, *Geometry*, 130.

20 Ibid.

21 Ibid.

22 Michel Serres, "Gnomon: The Beginnings of Geometry in Greece," in M. Serres (ed.), *A History of Scientific Thought* (Cambridge: Blackwell, [1989] 1995), 73–123. Cf. also Thomas L. Heath, in his commentary on the translation of the second definition of Euclid's second book, he describes it as "a thing enabling something to be known, observed or verified." Euclid, *Euclid's Elements*, trans. Thomas L. Heath (New York: Dover, [1908] 1956).

23 Serres, "*Gnomon*," 83–4.

24 Ibid., 80.

25 Serres, *Geometry*, 133.

26 Ibid., 134.

27 Ibid.

28 This is a term from algebraic topology. For a first acquaintance with the mathematical context of this term, cf. the *Wikipedia* entry: https://en.wikipedia.org/wiki/Covering_space (accessed January 7, 2018).

29 Serres, "Structure et Importation," 32.

30 Ibid., 34.

31 From the Latin *sensus*, from the PIE root **sent-*, "to go" (source also of Old High German *sinnan*, "to go, travel, strive after, have in mind, perceive"), cf. https://www.etymonline.com/word/sense.

32 Chirality is a key figure Serres develops in his book *Le Gaucher Boiteux*, and also in *L'Hermaphrodite*.

33 Cf. Chapter 2, "Quantum Literacy."

34 Serres, *Atlas*, 98; cf. the entire chapter, "Temps du Monde," 87–118.

35 Serres, *Geometry*, xxxiv (Emphasis in the original).

36 Ibid., xiv.

37 Ibid., 159.

38 Ibid., 134.

39 Ibid., 159.

40 Ibid., 157, "The information it shows or gives corresponds to its form and varies with it. According to the form, the information changes. Knowledge lies in the form."

41 Ibid., 158.

42 Serres, *L'Incandescent*, 298–302.

43 Ibid..

44 Ibid., 298, "Il faut donc revenir sur le devenir. Soit un état donné d'un système en évolution; je ne fais aucune hypothèse sur sa matière: solide ou fluide, liquide ou gazeuse, vivante ou culturelle, sociale ou langagière, virtuelle, mythique, artistique, n'importe, subjective même. Il change, il évolue selon la durée. Le processuel suit rarement une ligne unique, mais se développe plutôt en un espace-temps à plusieurs dimensions. Il peut y rencontrer une circonstance qui le fasse bifurquer: barrage dur, affluent visqueux, tension électrique, chaleur, froid, microbe, obstacle ou adjuvant inattendus, transformation de l'environnement, occasion que les Grecs nommaient kairos, coup de foudre, tonnerre ou amour, nez de Cléopâtre. La circonstance en question restructure l'état courant."

45 Serres, *Geometry*, 158.

46 Ibid., 159.

47 Ibid., xiv.

48 Serres, *Geometry*, 56.

49 Ibid., 47.

50 Ibid.

51 Ibid.

52 Ibid., 124.

53 Ibid.

4 Chronopedia II: Treasuring Time

1 Serres, "Ce que Thalès a vu," 175; "Thalès, lisant et relevant les traces du volume, ne déchiffre nul secret sinon celui de l'impuissance à pénétrer les arcanes du solide, où le savoir est à jamais enseveli, d'où l'histoire infinie des progrès analytiques jaillit comme d'une source."

2 Ibid.

3 Ibid., 226.

4 Serres, "Gnomon," 88–9.

5 Ibid., 88.

6 Ibid., 77.

7 Ibid., "All the cultural hegemonies of the world are impotent against this community and against the universality of this teaching."

8 From *ex-*, "out," + *serere*, "attach, join."

9 Serres, "Gnomon," 80.

10 Ibid.

11 Ibid., 86, my translation here deviates from the proposed one, which suggests the following: "The world lends itself to be seen by the world that sees it: that is the meaning of the word theory. Even better: a thing—the gnomon—intervenes in the world so that it might read on itself the writing it traces on itself. A pocket or fold of knowledge."

12 Ibid., 86–7.

13 Ibid., 86. There see footnote 10: "Algorithm: contrary to appearances, the word does not come from Greek but from Arabic and means a finite set of elementary operations for a computational procedure or the resolution of a problem."

14 In these descriptions, I follow mainly the account given by James Ritter in his essay "Babylon," in Serres (ed.), *History of Scientific Thought* (Cambridge: Blackwell, [1989] 1995), 17–43, as well as "Measure for Measure," in Serres (ed.), *History of Scientific Thought* (Cambridge: Blackwell, [1989] 1995), 44–72, both in Serres (ed.), *History of Scientific Thought.*

15 Ritter, "Measure for Measure," 62.

16 Ibid., 69, Ritter cites from the Rhind Papyrus.

17 Ibid., 96.

18 Ibid.

19 Serres, "Ce que Thalès a vu," 167.

20 Ibid., "Thalès arête le temps pour mesurer l'espace. Arrête la course du soleil a l'instant singulier des triangles isocèles, homogénéise la journée pour le cas général. … il faut geler le temps pour concevoir la géométrie."

21 See Michel Serres, *The Birth of Physics*, trans. David Webb and William Ross (London: Rowman and Littlefield, [1977] 2018).

22 Serres, "Ce que Thalès a vu," 164, "La pyramide est inaccessible, il invente l'échelle."

23 Ibid., 167. As Serres literally puts it, "Echanger les fonctions du variable et de l'invariable. L'origine de la géométrie est un confluent des genèses. A suivre les autres affluents," and more specifically he elaborates on "une genèse pratique," "une genèse sensorielle," "une genèse civil ou épistémologique," "une genèse conceptuelle ou esthétique."

24 Ibid., 167–8.

25 In "Gnomon," Serres writes, "And so it does not appear that the Ancients sought or thought of elements absolutely first or last: there are elements everywhere, in local tables" (112). He explains, "The term Elements, which translates into Latin and our modern languages the title used by Euclid and probably Hippocratus of Chios before him, originates from the letters L, M, N, in the same way as the alphabet spells the first Greek letters: alpha, beta, and the sol-fa sings the notes: sol, fa. The authentic title Stoicheia does indeed mean letters, understood as elements of the syllable or of the world" (111). And, further: "Again, what is an element? This mark, this line, the dash, the hyphen, in general the note, as these words were used by Leibniz. And in the plural, a series of these notes, a series generally grouped in a table or a chart of points and lines, in well-ordered lines and columns. As far as I know, the Elements of geometry also consisted of points and lines that we have to learn to draw. Today, as in the past, everywhere we see similar tables: the letters of the different alphabets,

numbers in all bases, axioms, simple bodies, the planets, markings in the sky, forces and corpuscles, the functions of truth, amino acids … Our memory preserves them so easily that they themselves constitute a memory in the triple sense of history—hence the commentaries—of automation and of algorithms" (113).

26 Serres, "Ce que Thalès a vu," 167.

27 Ibid., "organiser la représentation visuelle de ce qui défie le toucher."

28 Ibid., "partir de l'astronomie, inverser la question du gnomon."

29 Ibid., "gommer le temps pour métriser l'espace. Echanger les fonctions du variable et de l'invariable."

30 Ibid., 168, "Il s'agit moins d'un récit de constitution que de la mise en scène d'une forme déjà la."

31 A transport that requires an importation by means of a structure. Cf. ibid., 169, "Seul, le décryptage mathématique du texte peut faire voir le rapport du schéma qu'il implique à la mobilisation qui le transforme en récit transmissible."

32 Ibid., 167–70.

33 Serres, "Ce que Thalès a vu," 168, "Où est le point de vue? N'importe où. A la source de lumière. L'application, le rapport, la mesure sont possibles par un alignement d'amers: on peut voir alignés le Soleil et le haut du tombeau, ou le sommet de la pyramide et le point extrême de l'ombre portée. C'est dire que le site peut se déplacer."

34 Ibid., "Où est l'objet? Il faut qu'il soit, lui aussi transportable. De fait, il l'est: ou par l'ombre qu'il porte, ou par le modèle qui l'imite."

35 Ibid., "Où est la source de lumière? Elle varie, c'est le cas du gnomon. Elle transporte l'objet sous forme d'ombre. Elle se trouve dans l'objet, c'est ce que nous allons nommer le miracle."

36 Michel Serres, *L' Hermaphrodite* (Paris: Flammarion, 1987). Here, Serres elaborates on how the hermaphrodite is the offspring of Hermes and Aphrodite, impersonating the mixed origin of a "confluence," the origin of him who is never at rest (Hermes) and her, whose body is anadyomene, foam born and virtually excitable because distributed in the oceans.

37 Serres, "Ce que Thalès a vu," 173, "Le théâtre de la mesure montre le décodage d'un secret, le déchiffrement d'une écriture, la lecture d'un dessin."

38 Ibid., 221.

39 Ibid., 169.

40 Ibid.

41 Ibid., 171, "La géométrie de Thales exprime, dans la légende, le rapport entre deux aveuglements, celui de résultat de la praxis et celui de sujet de la praxis. Elle le dit et mesure le problème elle ne le résout pas, elle dramatise son concept et ne l'explique pas, elle désigne admirablement la question sans y répondre, elle raconte le rapport de deux chiffres, celui de maçon et celui de l'Edifice, sans les décrypter chacun."

42 Serres, "Ce que Thalès a vu," 170, "Quel est le statut du savoir implicite à une technique? Une technique est toujours une pratique enveloppant une théorie. Toute la question—ici la question d'origine—se résume en une interrogation sur le mode, sur la modalité de cet enveloppement. Si les mathématiques émergent un jour de certaines techniques, c'est sans doute par explicitation de ce savoir implicite. Qu'il y ait un thème du secret dans la tradition artisanale, cela signifie sans doute que ce secret est un secret pour tous, y compris pour le maitre. Il y a un savoir clair qui se cache dans la relation ouvrière aux pierres et aux moellons. Il y est caché, il y est enfermé, verrouillé à double tour: il y est à l'ombre. Il est à l'ombre de la pyramide. Voici la scène du savoir, la mise en scène de l'origine possible, rêvée, conceptualisée. Le secret du constructeur

et du tailleur de pierre, secret pour lui, pour Thales, et pour nous, et la scène de l'ombre. Sous l'ombre des Pyramides, Thales est dans l'implicite du savoir, que le Soleil, derrière, doit expliciter, en notre absence. Dès lors, toute la question du rapport entre le schéma et l'histoire, du rapport entre le savoir implicite et la pratique ouvrière, va être posée en termes d'ombre et le Soleil, dramatisation a la mode platonicienne, de l'implicite et de l'explicite, du savoir et de l'opération technique: le Soleil de la connaissance et du même, l'ombre de l'opinion, de l'empiricité, des objets."

43 Ibid., 179–80, "Si, par naissance de la géométrie, on entend l'émergence de l'absolue pureté sur l'océan nombreux de ces ombres, disons, quelques années après sa mort, qu'elle n'était jamais née. L'histoire des sciences mathématiques, dans sa continuité globale ou ses à-coups éclatants résout lentement et sans l'épuiser la question d'origine. Elle n'en finit pas d'y répondre et de s'en délivrer. Le récit d'inauguration est ce discours interminable que nous tenons sans repos depuis notre propre aurore. Qu'est-ce, au fait, qu'un discours interminable? Celui qui se rapporte d'un objet absent, d'un objet qui s'absente, inaccessiblement."

44 Serres, "Ce que Thalès a vu," 165, "La mesure, l'arpentage, directs ou immédiats, sont des opérations d'application. Au sens, bien sûr, où une métrique, une métrétique, ressortit à une science appliquée. Au sens où le plus souvent, la mesure est l'essentiel de l'application. Mais surtout au sens du toucher. Telle unité ou telle règle est appliquée sur la chose à mesurer, elle est plaquée dessus, elle la touche; ceci, autant de fois qu'il est besoin. La mesure immédiate ou directe est possible ou impossible autant que ce placage et possible ou ne l'est pas. Ainsi, l'inaccessible et ce que je ne puis toucher, ce vers quoi je ne puis transporter la règle, ce sur quoi l'unité ne peut être appliquée. Il faut, dit-on, passer de la pratique à la théorie, par une ruse de la raison, imaginer une substitute de ces longueurs ou mon corps ne peut accéder, la pyramide, le soleil, le navire à l'horizon, l'autre côté du fleuve. La mathématique serait le circuit des ruses."

45 Ibid., 175–6.

46 Ibid., "Sans doute, le vrai savoir des choses du monde gît dans l'ombre essentielle du solide, dans sa compacité opaque et noire, verrouillé à jamais derrière les multiples portes de ses bord, seuls attaqués par la pratique et par la théorie."

47 Serres, "Ce que Thalès a vu," 175–6.

48 Ibid., 178–9; "Or, justement, quand va mourir cette géométrie pure héritée du platonisme, quand nul ne va plus pouvoir se fonder sur l'intuition, quand va se fermer le théâtre de la représentation, le secret, l'ombre, l'implication va de nouveau exploser, parmi ces formes abstraites, aux yeux des mathématiciens étonnés—explosions annoncées avant tous ces décès tout au long de l'histoire. La droite, le plan, le volume, leurs intervalles et leurs régions, vont être reconnus chaotiques, denses, compacts … fourmillant à nouveau de replis et de cachettes noires. Les formes pures et simples ne sont ni si simples ni si pures, elles ne sont plus des connus théoriques complets, des vus et sus sans résidus, mais des insus théoriques objectifs infiniment répliques, d'énormes virtualités de noèmes, comme les pierres et les objets du monde, comme nos constructions de pierre et nos objets ouvrés. La forme cache sous sa forme des noyaux de savoir transfinis dont on se prend à craindre que l'histoire ne va pas suffire a les épuiser, des instances fortement inaccessibles comme des taches. Le réalisme mathématique s'alourdit et reprend une compacité qu'avait dissoute le soleil platonicien. Les idéalités pures ou abstraites font de l'ombre, sont pleines d'ombre, redeviennent noires comme la Pyramide ; la mathématique d'aujourd'hui, quoique maximalement abstraite et pure, se développe dans un lexique issu, pour partie, de la technologie. Nouvelle manière de réécouter la vieille légende égyptienne de Thales."

5 Banking Universality: The Magnitudes of Ageing

1 Serres, *The Incandescent*, 67.
2 Serres, "Ce que Thalès a vu," 175–6.
3 Serres, *The Incandescent*, 120.
4 Serres, *The Natural Contract*, 24, 90.
5 Serres, *The Incandescent*, 64-66. "White concepts form a group rather than a simple class: they proceed from each other. Seek freedom, and you will know; seek knowledge, and you will invent; seek knowledge and invention at the same time, and you will not be able not to love." He goes as far as to answer his own question, what metaphysics is useful for ("A quoi sert la Metaphysique?"), by speaking of this group of concepts as the incarnation of a generic "body" that is born from any body's body: "From the deprogrammed body set sail pagus and house—objective; next, from the relational set sail money and signs, to return in a loop to the subjective; then, individual and swept along, the body leaves again, in turn, for other externalizations, to the temple, the marketplace, the tribunal—collective—to all professions or kinds of work—objective—to the cognitive: symbol, apeiron …, lastly to the freedom that sums up and starts the loop again. Thus this class is structured as a group" (64). It is crucial for understanding why his book both is, and is not, a book on humanism: it is not a humanism that was sophistic; rather, it is one of sophistication—oscillating between truth (philosophy, ontology) and use/interest (sophists, humanism).
6 Ibid.
7 Ibid., 67, "Metaphysics serves to remain human and not die from it."
8 Serres, *Pantopie*, 300, "Messieurs, vous êtes mathématiciens, physiciens, astronomes, biologistes, etc., vous m'avez appris tout ce que je sais, mais, au fond du fond, vous ne m'avez appris qu'une chose. Cette chose, toute nouvelle et précieuse, vous l'avez tous inventée presque en même temps! Et c'est précisément ce qui vous fédère aujourd'hui et me permet de m'adresser à vous comme a une seule personne, comme à un singulier collectif. Aujourd'hui, pour la première fois dans l'histoire, toutes les sciences ont appris à dater leur objet. L'astronome sait dater précisément le big bang, les galaxies; naine blanche ou géante rouge ont une date de naissance précise; le biologiste sait dater la naissance des espèces aussi bien que des microbes; le géophysicien sait dater l'âge de la Terre; le paléoanthropologue sait dater chacun des hominidés qui sont apparus avant Homo sapiens; le linguiste sait dater la naissance de chacune des langues qu'il étudie, etc. … Je vous vois, et je ne vois qu'une seule et même personne. Vous.' Et, tout d'un coup, le spectre total de l'histoire s'est déployé devant nous."
9 Serres, "Vie, information, deuxième principe," 42–3.
10 Serres, *The Incandescent*, 37, "Like a temporal transcendental, certain conditions for knowing date from hundreds of millions of years ago."
11 Michel Serres, "Moteurs, suivi de régression," in M. Serres (ed.), *Hermes IV, La Distribution* (Paris: Les Éditions de Minuit, 1977), 41–86, here 82.
12 Cf. Chapter 2, "Quantum Literacy," here "Taking Ignorance into Account: Quantifying Strangeness."
13 Serres, "Moteurs," 90, "On bloque alors la fluence du transitif, la référence à la situation dans le cours du processus, et on atteint la stabilité, l'invariance: toujours. Enfin, l'utilisation du l'est n'indique pas autre chose que l'invariant."
14 Serres, "Trahison: la thanatocratie."
15 Serres, "Vie, information, deuxième principe," 44.

16 Cf. Giuseppe Longo's discussion on the concept of invariance in his article "Synthetic Philosophy of Mathematics and Natural Sciences."

17 Cf. Anne Crahay, *Michel Serres, la mutation du cogito: Genèse du transcendantal objectif.*

18 Michel Serres, "Ce qui est écrit sur la table rase," in M. Serres (ed.), *Hermes II, L'Interférence* (Paris: Les Éditions de Minuit, 1972), 106, "Voici donc la variation généralisée, la variation mondiale … existent comme supports objectifs d'une information, qu'ils reçoivent, conservent et rendent."

19 Serres, *The Incandescent*, 37, although here my own translation deviates a little from that by Randolph Burks, who omits to mention "material" as set apart from "software" and "hardware" by translating "This type of intrahardware software conditions our cognitive performances, like a kind of objective transcendental." The original French is: "Ce type de logiciel intramatériel conditionne nos performance cognitives, comme une sorte de transcendantal objectif" (61).

20 Michel Serres, "Information and Thinking," in Rick Dolphjin and Rosi Braidotti (eds.), *Philosophy after Nature* (London and New York: Rowman and Littlefield, 2017), 13.

21 Serres, "Introduction," in M. Serres (ed.), *Hermès I, Communication* (Paris: Les Éditions de Minuit, 1968), 10, "La mathématique n'est plus un support, ou un garde-fou, elle est un dictionnaire."

22 Serres, "Ce qui est écrit sur la table rase," in M. Serres, *Hermès II, L'Interférence* (Paris: Éditions de Minuit, 1972), 110. The expression "silent spectral sculpture" is not literally used by Serres.

23 According to this thinking, Serres figures out an entire theory of how to think about waste and pollution: whenever one pole of the equipollence (between rationality or reality) dominates the other, that is when the identity element is not adequate to in principle (however temporarily) accommodate all without rest, it produces pollution. Cf. Serres, *The Natural Contract*, 24–5, "If our rational could wed the real, the real the rational, our reasoned undertakings would leave no residue; so if garbage proliferates in the gap between them, it's because that gap produces pollution, which fills in the distance between the rational and real." Cf. also Michel Serres, *Le Mal propre. Polluer pour s'approprier?* (Paris: Le Pommier, 2008), as well as Serres, *Pantopie*, 233.

24 Michel Serres, "Interview de Michel Serres sur l'autorité," APEL the Association of Parents of School Kids in France, in the context of a Conference entitled Autoriser l'autorité in Montpellier 2010, available online: https://www.youtube.com/watch?v=GeJgTmZ9EBc (accessed January 2, 2018).

25 Michel Serres, *Hermes V, Le passage du nord-ouest* (Paris: Éditions de Minuit, 1980), here "Espace et temps," 67–83, and "Histoire: L'Univers et le lieu. Obstruction," 84–92.

26 Serres, *The Incandescent*, 62.

27 Cf. Serres, *Genesis.*

28 An old Latin word for the loftiness in thought or purpose, which Serres, however, does not use (to my knowledge).

29 Cf. Serres, *The Incandescent*, chapter "Descent into dedifference," 40–49.

30 Ibid., 65, "is there a bound, a lower limit for dedifferentiation?"

31 Ibid., 64, "Let's call metaphysics the discipline that deals with the group of white concepts. Universals."

32 Cf. Serres, *Hermes V, Le passage du nord-ouest* (Paris: Éditions de Minuit, 1980) and Serres, *Genesis.*

33 Zalamea, *Synthetic Philosophy, Contemporary Mathematics*, and Longo, "Synthetic Philosophy of Mathematics and Natural Sciences."

34 Cf. especially Elias Zafiris, *Natural Communication. The Obstacle-Embracing Art of Abstract Gnomonics* (Basel/Vienna: Birkhäuser, forthcoming).

35 Louis Hjelmslev, *Prolegomena to a Theory of Language* (Madison: University of Wisconsin, [1943] 1969). Cf. also Nikola Marincic's study on Louis Hijelmslev's glossematics in relation to thinking about coding as literacy: *Towards Communication in CAAD.*

36 Whitehead, *A Treatise on Universal Algebra.*

37 The Church-Turing paradigm establishes a specific number class, called "computable numbers" which grants one dominant and homogeneous arithmetics. Turing himself overcame this paradigm with his attention to morphogenesis. Cf. Turing, "On Computable Numbers," and "On the Chemical Basis of Morphogenesis."

38 Hjelmslev, *Prolegomena*, 52.

39 This is indeed, I think, why Hjelmslev was so much concerned with pointing out that his approach to language is one that affords a *science* of language; that its interest concerns languages' generic objectivity rather than the subjective plays of interpretation it also affords, or the imperialist claims to how proximate languages are with regard to a supposedly original language.

40 The main aim of Hjelmslev was precisely this—to do away with the necessity of a metaphysics for science, cf. the last chapter in his *Prolegomena*,

41 Identity, for Serres, is a question of "appartenance," belonging-with, affiliation, German *Zugehörigkeit*. Cf. Serres, *The Incandescent*, the chapter "Identity, belongingnesses," 71–85.

42 Cf. Serres, "Revisiting the Natural Contract," in Arthur and Marilouise Kroker (eds.), *1000 Days of Theory*, 2006. Available online: http://ctheory.net/ctheory_wp/revisiting-the-natural-contract/ (accessed January 12, 2018).

43 Serres, *The Incandescent*, 120.

44 Ibid., 30.

45 Ibid., 31, "All things, in principle, act as memories. The universe banks accounts. All things are of number, the memory of the world conserves traces." Here my own translation deviates from that by Randolph Burks, who translates: "All things, in principle, act as memories. The bank-Universe contains accounts. All things are numbers; the memory-world preserves traces." The French original is: "Toutes choses, en principe, se comportent comme des mémoires. L'Univers banque des comptes. Toutes choses sont nombres, le monde mémoire conserve des traces" (53).

46 I developed this argument on Serres's reading of Balzac's *La Belle Noiseuse* in: Vera Bühlmann, "'Ichnography'—The Nude and Its Model. The Alphabetic Absolute and Storytelling in the Grammatical Case of the Cryptographic Locative."

47 Serres, *The Incandescent,* 10, "We are almost all as old as the Earth." The original French says it more dramatically: "nous voilà tous presque aussi vieux que la Terre" (21).

48 Ibid., 7.

49 Ibid., 10–11.

50 Ibid., 36.

51 Ibid., 31, although here my own translation deviates from that by Randolph Burks, cf. footnote 45.

52 Michel Serres, *Les Nouvelles du Monde* (Paris: Flammarion, 1997).

53 Cf. Serres, *The Incandescent*, here the chapter "An Admission by Identity: Entirely Innate, Entirely Acquired," 76–7, where it reads, "There is no discussion or contradiction or even proportion between the acquired and the innate, between the two cards I'm talking about: entirely innate, entirely acquired, humanity is formed

by this strange addition," and further: "In my body, my soul and my palimpsest-understanding, a thousand texts and drawings arrange to meet each other, heavily overburdened, speedily forgotten, memorized, overlapping, ceaselessly erased and nonetheless always repainted and rewritten in reshaped furrows. Everything written on this absence; or: no one plus the others; that's the self. *Ego nemo et alii*" (77).

6 The Incandescent Paraclete: Tables of Plenty

1   Michel Serres, *Hominescence* (Paris: Le Pommier, 2001), 14.

2   Serres, *The Natural Contract*, 92.

3   Serres, *The Parasite*, 88.

4   Serres, *Genesis*, 9.

5   Ibid., 8.

6   Serres, *Rome*, 18.

7   Serres, *Genesis*, 1.

8   Ibid.

9   Ibid.

10 Ibid., 5.

11 Ibid.

12 Ibid.

13 Serres, *Rome*, 1.

14 Serres, *The Parasite*, 41.

15 Serres, *The Incandescent*, 145.

16 Cf. Georg Wilhelm Friedrich Hegel, *Lectures on the Philosophy of History*, trans. J. Sibree (New York: Prometheus, 1872–1900); Introduction, "Classification of Historic Data," 109–10. Also: Jacques Derrida, "White Mythology: Metaphor in the Text of Philosophy," trans. F. C. T. Moore, *New Literary History*, 6 (1), "On Metaphor" (John Hopkins University Press, Autumn, 1974): 5–74.

17 Serres, *Rome*, 1.

18 Serres, *The Incandescent*, 145, "The modern West [L'Occident moderne] benefited from a rare bit of luck. Little by little, then en masse, it converted to an imported religion, born on another land than its own, the Holy Land, located elsewhere, in Palestine, where holy history unfolded, the history of the chosen people and of the redemption by the come Messiah. The decisive events for salvation took place elsewhere than in its home and concern other personages than the gods of its culture. Neither the holy geography nor the holy history unfold beneath its feet, the persons concerned not speaking its language. In no longer deifying their own soil, Western peoples thus separated the spiritual from their native roots. Thus they left the *pagus* of paganism, the little gods of their fields. The West has carried out said deterritorialization from its origin. Or rather: it owes its sudden bifurcation to this coming unstuck with respect to the soil."

19 Serres, *Rome*, 1.

20 Ibid., 18.

21 Ibid., 5.

22 Serres, *The Incandescent*, 145.

23 Serres, *Rome*, 18.

24 Ibid., 96.

25 Ibid., 216.

26 Ibid., 2.

27 Ibid., 20.

28 Serres, *The Incandescent*, 145.

29 Ibid.

30 Ibid.

31 Serres, *The Parasite*, 44.

32 Serres, *Rome*, 4.

33 Marcel Hénaff, *The Price of Truth. Gift, Money, and Philosophy*, trans. Jean-Louis Morhange (Stanford: Stanford University Press, 2010).

34 Serres, *Hominescence*, 14.

35 Ibid.

36 Serres, *The Incandescent*, 145.

37 Serres, *Darwin, Bonaparte et le Samaritain*, 31.

38 Wiener, *Cybernetics*.

39 Serres, *Darwin, Bonaparte et le Samaritain*, 28, "Nouvelle, assurément, la science universelle d'aujourd'hui pourrait se nommer ichnographie, puisque les sciences dites dures interprète les empreintes laissées par les évènements des mondes inertes et vivants—*ichnos*, en greque, "la trace de pas"—comme la philologie, l'histoire, bref les sciences dite douces interprètent les restes et les écritures humain. Ce terme d'ichnographie désignait au XVIIème siècle l'intégrale des profils sous lesquels on peut voir un objet."

40 Ibid., 163, "L'histoire commence avec l'écriture. Or, comme nous, tout choses du monde laissent des traces sur toute choses et sur nous. Cette assertion acquiert donc, on l'a compris, une portée universelle. De même que, pour comprendre un message écrit ordinaire, il convient d'apprendre la forme et l'association des lettres laissées par le poinçon, le style et l'encre sur un support, de même les sciences, peu a peu, découvrent les codes sous lesquelles gisent mille et un sens que cache et révèle le lueur de l'univers, la radioactivité, le climat, les fossiles."

41 Ibid., "Galilée disait le monde écrit en langue mathématiques; mieux vaudrait le dire code; tel équation du deuxième dégrée code la chute des corps comme tel contrat écrit pour une vente ou un mariage. Juridiques ou physique, tout lois sont, en même titre, des codes. Ecrits, l'Univers, ainsi caché ou protégé, git sous mille codes que déchiffrent mille choses et que nous déchiffrons, comme nous faisons pour les messages que nous échangeons en milles langues humains. Dense de sens, le monde entre dans l'histoire."

42 Ibid., 31, "Cette reconstruction de la connaissance pose, pour finir, une question proprement métaphysique: existe-t-il un porte empreinte, cette mystérieuse *chôra*, dont le Timée de Platon évoque le caractère universel, le comparent à l'utérus maternel? Peut-on concevoir, comme support universel, l'espace-temps que je m'apprête à décrire, table rase sur laquelle toutes les choses, tous les vivants du monde, toutes les langues humains laissent leurs traces?"

43 www.etymonline.com; modern for "now existing"; 1580s, "of or pertaining to present or recent times"; from the Middle French *moderne* (15c.) and directly from the Late Latin *modernus*, "modern" (Priscian, Cassiodorus); from the Latin *modo*, "just now, in a (certain) manner," from *modo* (adv.), "to the measure," ablative of *modus*, "manner, measure."

44 Serres, *Geometry*, xiii.

45 Serres, "Ce que Thalès a vu," 178.

46 Michel Serres, "Les Nouvelles Technologies: Révolution Culturelle et Cognitive" (Conférence de Michel Serres for the 40th anniversary of l'INRIA, Paris, 2007).

Available online: https://interstices.info/jcms/c_33030/les-nouvelles-technologies-revolution-culturelle-et-cognitive (accessed January 12, 2018).

47 Serres, *The Parasite*, 40.

48 Peter Sloterdijk, *Tau von den Bermudas. Uber einige Regimes der Einbildungskraft* (Frankfurt am Main: Suhrkamp, 2001).

49 Ibid., 41.

50 Michel Serres, "Le réseau de communication: Pénélope," in M. Serres (ed.), *Hermes I, La Communication* (Paris: Éditions de Minuit, 1968), 11–35.

51 Serres, *The Parasite*, 42.

52 In that it can be experienced. As a concept, he is very clear, the universal is "without version," it is that which is "not communicable"; as such, it is what makes the immediacy of experiences possible: experiences cannot be exhaustively represented in any communication because there is always something universal in it—if it is a genuine experience, a euphoric one, one that puts us in touch with things, with the world. Cf. *Hermes I, La Communication*, 10.

53 Cf. Serres, *Darwin, Bonaparte et le Samaritain*, 168, cf. chapter "Reseaux, percolation, paysage."

54 Serres, *The Five Senses*, 343.

55 Ibid., 345.

56 Ibid., 213.

57 Serres, *Statues*, 51.

58 Serres, "Information and Thinking," in Rick Dolphjin and Rosi Braidotti (eds.), *Philosophy after Nature* (London and New York: Rowman and Littlefield, 2017), 13–20, here 17.

59 Serres, *Darwin, Bonaparte et le Samaritain*, 7, "Les historiens se vantent volontiers d'exercer, de pratiquer, de célébrer la mémoire, alors que leur discipline se définit plutôt comme une série d'oubli."

60 Ibid., 7–8.

61 Ibid., 9, "Autrement dit, l'humain constitue le sujet exclusive ou central de l' histoire, de l'ethnologie et de la préhistoire, voire leur références. Avec une complaisance pathétique, nous nous mirons, narcisses, en ces trois disciplines."

62 Cf. also Jean-François and Michel Serres, *Solitude: Dialogue sur l'engagement* (Paris: Le Pommier, 2015).

63 Serres, *Darwin, Bonaparte et le Samaritain*, 21, "la question, traditionnelle en philosophie du sens de l'histoire humaine se généralise-t-elle au Grand Récit qui la précède et l'accompagne?"

64 Ibid., 23, "Autrement dit, ce récit n'a de sens que si on le raconte au futur antérieur, chemin que je viens, justement, d'emprunter, alors qu'il reste sans finalité d'amont en aval. Cette absence de finalité est l'outil de base des savants: si elle existait, ceux-là ne feraient plus de science."

65 Ibid., 25.

66 Ibid., 25–6, "Cela signifie, je le souligne de nouveau, que, considérée à partir de l'amont, les émergences ne sauraient se prévoir, mais que, vue de l'aval, elle peuvent apparaitre comme rationnelles, découlant d'une ou plusieurs causes, ou plutôt conditions, que l'expert découvre comme nécessaires mais qui n'accèdent jamais à la suffisance. Cette distinction logique et mathématique éclaire de manière rigoureuse ce mouvement rétrograde du vrai. Avant, aveuglement; par après, diverses variétés de clairvoyance. J'aime à le redire, le Grand Récit est bien un récit, roman scandé de suspenses, tragédie ou comédie, traversé de messagers apportant des nouvelles qui, en faisant bifurquer le temps, augmentent la tension de l'attente, comme dans l'histoire ordinaire."

67 Ibid., 23, "Multimilliardaire, le Grand Récit peut-il passer pour l'âge du Père?"
68 Ibid., "dans le sens inverse, en remontant vers le commencement."
69 https://www.etymonline.com/search?q=parasite.
70 Serres, *The Parasite*, 45.
71 Serres, *L'Incandescent*, especially 54–6.
72 Ibid., 53, "Toutes choses, en principe, se comportent comme des mémoires. L'Univers banque contient des comptes. Toutes choses sont nombres, le monde mémoire conserve des traces."
73 Serres, *Rome*, 31.
74 Serres, "Information and Thinking," 17–18.
75 Serres, *The Parasite*, 43.
76 Michel Serres, "La confession fraternelle," *Empan* 48 (2002): 11–16. Here cited from the unpublished translation by Kris Pender, "Fraternal Confession." Available online: https://www.academia.edu/11074066/Michel_Serres_-_Fraternal_Confession (accessed January 5, 2018).
77 Michel Serres, "Noise," trans. Lawrence R. Schehr, *SubStance* 12, no. 3 (1983): 48–60, here 55. The article makes up the first chapter of Serres's 1982 book *Genesis*.
78 Online Etymology Dictionary: "crypt." Available online: http://www.etymonline.com/index.php?term=crypt&allowed_in_frame=0 (accessed January 5, 2018).
79 Serres, "Fraternal Confession," 8.
80 Ibid., 7. The French original is available online: https://www.cairn.info/revue-empan-2002-4-page-11.htm (accessed January 12, 2018). Here it reads: "L'Évangile appelle Satan le Maître du monde ; il vous emmène sur une très haute montagne, vous montre tous les royaumes dans toute leur gloire et promet de vous les donner à la condition que vous vous prosterniez devant lui. Désobéissez-lui donc, sans aucune condition."
81 Serres, "Fraternal Confession," 7.
82 Ibid., 8.
83 Ibid., 6.
84 Ibid.
85 Ibid., 7.
86 Ibid.
87 Ibid.
88 Ibid., 1.
89 This very dynamic is the theme of Serres's article "Moteurs, suivi de régression." Its neglect is what Serres discusses as thanatocracy in "Trahison: la thanatocratie."
90 Serres, *The Five Senses*, 262.
91 Serres, *Statues*, 51.
92 Ibid.
93 Ibid.
94 This idea builds on Serres's problematization of the universal quantifier (All) and the positive methodology it triggers in history, where it is concerned with operations of unification, integration and mean averages. Serres criticizes the idea of seeing in history the complementary pole to the rationalism of an empirical scientist, whose methods are geared towards exposing an object, towards setting a particular domain apart from the whole, and hence work negatively by operations of exclusion, subtraction and elimination. Serres's point is that the universal ought to be considered "under the form of distribution," and "in the sign of quantification." Cf. Michel Serres, "Histoire: L'Univers et le lieu. Obstruction," in M. Serres (ed.), *Hermes V, Le passage du nord-ouest* (Paris: Éditions de Minuit, 1980), 84.

95 Serres gives a surprising twist to the Cartesian skepsis: *dubitare* comes from *duo-habere*, he maintains, for being undecided about what something is to be taken for. Serres twist is in regarding this moment as what he calls "a frequentativum," as a returning oscillation between two positions. Cf. Michel Serres, "Solides, Fluides, Flames," in M. Serres (ed.), *Hermes V, Le Passage Nord-Ouest* (Paris: Les Éditions de Minuit, 1980), 40–66, here 41.

96 Cf. Serres, "Vie, information, deuxième principe," 45.

97 I have elaborated on this in my articles "Maxwell's Demon (Non-Anthropocentric Cognition)" and "Negentropy," in Rosi Braidotti and Maria Hlavajova (eds.), *Posthuman Glossary* (London: Bloomsbury, 2018). Cf. also Chapter 2 in the present volume, "Quantum literacy."

98 Cf. Serres, "Vie, information, deuxième principe," 46, "Bref, étrange est improbable, rare, et le calcul des miracles donne des nombres au plus près voisinage de zéro. D'où Borel: les phénomènes les plus improbables ne se produisent jamais. D'où Monod: les phénomènes les plus improbables, lorsqu'ils existent, et reproduisent; ils se produisent parce qu'ils se reproduisent; l la grande question du hasard et de l'entropie se dessine déjà. Et donc l'étrangeté se quantifie en précision: à haut niveau d'information ou de néguentropie. Étrange est l'écho du hasard et l'objet celui d'une nécessité. L'adjectif ne qualifie pas le nom, il le quantifie." My own translation: "Put briefly: strange means improbable, rare, and the calculus of miracles yields numbers that stand in closest proximity to zero. Hence Borel: the most improbable phenomena are not ever yielded in this calculus. Hence Monod: When highly improbable phenomena do happen, then they reproduce themselves; they happen because they reproduce themselves. The grand question of chance and entropy already begins to show here. And thus strangeness receives a precise quantification: on high niveau of information or negentropy. Strange is the echo of chance, and the object is one of necessity. The adjective [strange, VB] does not qualify the substantive, it quantifies it."

99 Serres, *Statues*, 5.

100 I cannot elaborate on this here. It is the complex issue at the core of Serres's foundations trilogy: the royal victim is one of the origins of geometry within Serres's mathematical realism. Cf. especially: *Geometry*, here "First in the Rite: The Royal Victim. Spaces of Exclusion: Political Origins."

101 Serres, *Birth of Physics*, position 3202.

102 Serres, "Vie, information, deuxième principe," 45.

103 Serres, *Statues*, 5.

104 Archimedes' *The Sand Reckoner* established a theory of how to address large numbers which does not seek to identify their in(de)finiteness, but which proceeds by encrypting it (their in(de)finiteness)—a procedure which finds wide application in today's informatics, under the name of "double recursion." The number system in Archimedes' time had identified as its largest number one that they called "myriad" (today 10'000, or 10^8), and Archimedes began to use this name in a recursive manner, by taking that name as a code (rather than as a proper name) which could be operationalized by declaring that it must be unit of this operative self-referentiality. He built different classes of numbers, myriads of myriads ($10^8 \times 10^8 = 10^{16}$) and myriads of myriads of myriads, and so on. Cf. Piergiorgio Odifreddi and S. Barry Cooper, "Recursive Functions," in Edward N. Zalta (ed.), *Stanford Encyclopedia of Philosophy* (Fall 2012 Edition). Available online: http://plato.stanford.edu/archives/fall2012/entries/recursive-functions (accessed January 12, 2018), especially the paragraph "Double Recursion." Serres

foregrounds the importance of Archimedes' theory in his *Birth of Physics in the Text of Lucretius*: "Atom-letters do not work like numbers (*chiffres*). Whatever the base of numeration, in fact, or the alphabet of the ciphering (*chiffrement*), the various combinations of these signs among themselves produce acceptable numbers. In this way, the interconnection of atoms in things, conjunction, is ciphered, nature is coded. Atomic physics discovered the key to the code. Now the cipher is hidden in its turn, since atoms, subliminal, are imperceptible and very great in number. That atoms are letters is a thesis that heralds the great classical philosophies, the idea of ciphering and the secret code, the global working of physical science. Now read Archimedes' *The Sand-Reckoner* and you will find a pre-combinative arithmetic that formalizes this idea. Physics is indeed an activity of deciphering or decoding." Serres, *Birth of Physics*, position 3242.

105 Here we have the crucial diversion Serres's thinking takes from that of the common reception of Leibniz's Universal Characteristics. This reception sees in the characteristics one basic "alphabet" that would underlie, in the sense of being a referent to, all alphabets. Serres's reading of Leibniz is one from the point of view of twentieth-century mathematics; it is structuralist; and it witnesses in Leibniz's "interest of a metaphysics that were mathematical" (position 138) his own position towards how to deal with the formal language of nineteenth- and twentieth-century symbolic and universal algebra. Serres, *Le système de Leibniz*, here cited from the Kindle edition; my own translations.

106 Ibid., Kindle edition: position 481.
107 Ibid., position 575.
108 Ibid., position 182.
109 Ibid., position 180 (emphasis mine).
110 Ibid., position 188.
111 Ibid., position 224.
112 Ibid.
113 Ibid.
114 Ibid., position 235, "Cela n'est encore qu'une approximation par schéma, mais elle emporte un secret que nous cherchions confusément."
115 We have here the old theme of the demonstrations of the existence of God by formal proofs. Cf. Jean-Yves Girard's three lectures at the Centre Henri Poincaré (2014) as a background discussion of how, also in mathematics, the formulation of the question predisposes the set of possible answers one can find, leaving the realist scientist with an irreducible uncertainty, even if the idealist scientist will have proved "certainty"; the first lecture is entitled: "Qu'est-ce qu'une réponse? (l'analytique)," the second lecture: "Qu'est-ce qu'une question? (le format)," the third lecture: "D'où vient la certitude? (l'épidictique)." Available online: https://www.youtube.com/watch?v=f7sT0J74pHI (accessed June 30, 2016).
116 Serres, *Le système de Leibniz*, Kindle edition: position 481.

7 Sophistication and Anamnesis: Remembering an Abundant Past

1 Hénaff, *Price of Truth*, 20.
2 Ibid.

3 Serres, "Vie, information, deuxième principe," and especially also "Mathématisation de l'empirisme" in *Hermes II, L'Interférence* (Paris: Les Éditions de Minuit, 1972), 195–200.
4 Monod, *Chance and Necessity*, 60–1, "The key to the riddle was provided by Leon Brillouin, drawing upon earlier work by Szilard: he demonstrated that the exercise of his cognitive function by the demon had to entail the consumption of a certain amount of energy which, on balance, precisely offset the lessening entropy within the system as a whole. So as to work the hatch 'intelligently', the demon must first have *measured* the speed of each particle of gas. Now any reckoning-that is to say, any acquisition of information-presupposes an interaction, in itself energy-consuming. This famous theorem is one of the sources of modern thinking regarding the equivalence between information and negative entropy. The theorem interests us here precisely because enzymes, at the microscopic level, exercise an order-creating function. But this creation of order, as we have seen, is not gratuitous; it comes about at the expense of a consumption of chemical potential. In short, the enzymes function exactly in the manner of Maxwell's demon corrected by Szilard and Brillouin, draining chemical potential into the processes chosen by the program of which they are the executors."
5 Hénaff, *Price of Truth*, 21.
6 Ibid., 5.
7 Michel Serres, "Mathématisation de l'empirisme," in M. Serres (ed.), *Hermes II, L'Interférence* (Paris: Les Éditions de Minuit, 1972).
8 Ibid.
9 Ibid., 20.
10 Ibid.
11 Ibid.
12 Ibid., 7.
13 Ibid.
14 Ibid., 8.
15 Ibid., 10.
16 Ibid., 11.
17 Ibid.
18 Serres, "Structure et importation," 43.
19 Ibid., 21, "Loin d´être la clé mystérieuse qu'on imagine ouvrir toutes les portes, elle n'est qu'une notion méthodique claire, distincte et lumineuse."
20 Ibid.
21 Ibid., "une définition normative, cathartique et purgative."
22 Especially: Gaston Bachelard, *Le nouvel ésprit scientifique* (Paris: Alcan, 1934); *La philosophie du non: essai d'une philosophie du nouvel esprit scientifique* (Paris: Presse Universitaire France, 1940).
23 Serres, "Structure et Importation," 22, "Par généralisation de l'idée classique du vrai et admission de la notion de sens, le romantisme est une tentative d'assomption et de promotion des contenus culturels comme tels; il introduit par là ce projet, sur lequel nous vivons encore, de comprendre le pluralisme des significations de décoder tous les langages qui ne sont pas forcément ceux de la raison pure."
24 Ibid., "Si le problème classique est celui de la vérité, le champ de ce problème la raison, le problème romantique est celui du sens et son champ l'ensemble historique des attitudes humaines; alors l'horizon méthodique du premier est celui de l'ordre (des déductions, des thèmes, des conditions, etc.), l'horizon méthodique du second est celui du symbole."

25 Ibid., 34–5, "Donner aux figures de cet autre monde dionysiaque des signification
    épaisse, compactes et obscures où se projettent l'âme humaine, son affectivité, son
    destin … Mais, outre qu'elles sont des symboles de l'histoire, ne seraient-elles pas dans
    leur ultime surcharge, dans leur dernière détermination, des modèle signifiants de
    structure transparente, de l'intellect, et de la science?"
26 Michel Serres, `Les anamnèses mathématiques', in *Hermes I, La Communication*,
    78–112 (Paris: Les Editions de Minuit, 1968).
27 Ibid., 91.
28 Ibid.
29 Ibid., 109–10.
30 Serres, "Structure et Importation," 22.
31 Ibid., 24.
32 Ibid., 22, "Dans le moment où on élargit le champ des questions, dès le moment ou
    l'obscurité du sens doit être assume comme telle, l'archétype de référence se trouve
    désadapte. Le domaine du sens ne mime plus aucun archétype rigoureux ou ordonne,
    aucun modèle ne tout arme de la pure raison. Il faut alors choisir un archétype dans le
    domaine même du sens et projeter sur ce modèle toute l'essence du contenu culturel
    analysé. Au lieu de faire référence à un modèle idéal comme à un index normatif,
    il faut construire un modèle concret à l'intérieur même du champ analysé et faire
    référence à son contenu plus qu'à son ordre. Tel contenu ne mime plus un modèle
    idéal, mais répète, contenu pour contenu, un symbole universel et concret."
33 Ibid., 22, "En ce temps-là les symboles sont descendus du ciel sur la terre; pas tout à
    fait cependant, car ils ne sont descendus que du ciel des idées sur la terre ou l'histoire
    des mythes."
34 Ibid., 22, as Serres adds in a footnote: "Où l'on voit que le pure déviant le mythique, ce
    dernier étant à la fois universel et singulier."
35 Ibid., 23, "En ce sens, la technique des analyses de Hegel, Nietzsche et Freud est
    symbolique et archétypique: toute la question est de savoir ou choisir l'archétype, dans
    quel ensemble symbolique puiser. En gros, les analyses symboliques du XIXe siècle
    choisissent leurs modèle dans l'histoire mythique: ainsi Apollon, Dionysos, Ariane,
    Zarathoustra, Electre, Oedipe, etc., représentent éminemment (symbolisent) la totalité
    de l'essence d'un contenu culturel de signification. Le sens de ce contenu est compris
    et assume dès que l'on montre qu'il recommence, qu'il réitère l'archétype, qu'il le
    réalise a nouveau, qu'il le fait passer du mythe a l'histoire, de l'éternel a l'évolutif."
36 Ibid., 23, "Du contenu a son symbole, il y a correspondance sens à sens, et cette
    correspondance engendre l'histoire ou le Retour éternel, si bien que la technique
    d'analyse symbolique est liée a la conception de l'histoire; inversement, les typologies
    historiques sont engendrées par l'ensemble d'archétype choisi."
37 Ibid., 23, "Du classicisme au romantisme, la notion de modèle passe du clair à l'obscur
    (c'est à dire, dans le champ des problèmes, du vrai aux signifiant), du normative aux
    symbolique, du transcendant a l'originel."
38 Ibid., 24, "Cette analyse rapide fait voir les notions sur lesquelles nous avons vécu
    jusqu'à hier: problème du sens et du signe, symbolisme et langages, archétypes et
    histoire, compréhension de contenus culturels obscurs, fascination de l'originel et de
    l'originaire et ainsi de suite."
39 Ibid., 24, "Mais ce qu'il faut souligner, c'est la variation des modèles choisis. Ce dont
    on n'a pas eu conscience, mais ce qui est pour nous désormais clair comme mille
    soleils, c'est qu'en variant sur nos problèmes nous avons varié sur nos références."
40 Ibid., 21.

41 Ibid., 24, "C'est que l'analyse symbolique du romantisme n'est pas un miracle méthodique original, mais une étape dans une variation."

42 Roland Barthes, "The Eiffel Tower," in Susan Sontag (ed.), *A Barthes Reader* (New York: Hill and Wang, 1983), 236.

43 Ibid., 236–7.

44 Barthes, "The Eiffel Tower," 237.

45 Ibid. (emphasis in the original).

46 Ibid., 238.

47 Ibid., 237 (emphasis in the original).

48 Ibid.

49 Ibid.

50 Ibid., 250.

51 Ibid., 241.

52 Ibid., 248.

53 Ibid., 239.

54 Ibid., 248.

55 Ibid., 249.

56 Ibid., 250.

57 Ibid.

58 Ibid.

59 Ibid.

60 Ibid.

61 Ibid., 241.

62 Ibid., 242 (emphasis in the original).

63 Ibid.

64 Ibid., 243.

65 Serres, "Structure and Import," 24.

66 Barthes, "The Eiffel Tower," 243.

67 Ibid.

68 Ibid.

69 Ibid.

70 Ibid., 240.

71 Ibid.

72 Ibid.

73 Ibid., 241.

74 Ibid., 237.

75 Ibid., 243.

76 Ibid., 244.

77 Ibid., 246.

78 Ibid.

79 Ibid.

80 Ibid., 247.

81 Ibid.

82 Ibid.

83 Ibid.

84 Ibid., 248.

85 Ibid.

86 Serres, "Structure et importation," 34.

87   Barthes, "The Eiffel Tower," 244.
88   Ibid.
89   Ibid., 247.
90   Ibid., 248.
91   Ibid., 242.
92   Ibid.
93   Ibid., 250.
94   Serres, *Le Tiers-Instruit*.
95   Serres, "Structure et importation," 34. Also: "Contrary to what is always said, science does not cancel out non-science … Myth remains dense in knowledge, and vice versa," in Michel Serres, *Feux et signaux de brume. Zola* (Paris: Grasset, 1975), 32, 49.
96   Ibid., 24: "Posez le problème du vrai, vous n'aurez que la mathématique comme main-courante limite, posez le problème de l'expérience, vous n'aurez que la mécanique, la physique ou philosophie de la nature."
97   Serres, *Statues*, 51.
98   Ibid., 50.
99   Serres, "Structure et importation," 24. Serres speaks of the double make-up of Bachelard's approach as a short circuit between "une psychanalyse de la connaissance objective" and "une psychanalyse de l'imagination matérielle signifiante."
100 Ibid.
101 Ibid., 25, "posez enfin celui de la signification des cultures, vous n'aurez que l'ensemble des archétypes fournis par la mémoire immémoriale de l'humanité."
102 Ibid., 26, "Toute question méthodique ou critique tourne, des lors, autour de la notion de sens; et, si j'ose dire, autour de sa quantification."
103 Ibid., "Soit une forme quelconque, à laquelle nous voulons assigner une quelconque fonction méthodique. Supposons que nous la remplissions de sens, que nous la chargions et surchargions de signification."
104 Ibid., "matérielle, historique, humaines, existentielle … jusque dans le précis de la singularité."
105 Ibid., "Cette forme devient alors un archétype, c'est à dire la référence d'une analyse symbolique: le langage du sens ne possède comme termes que des archétypes, le langage du sens n'est dicible qu'en idéogrammes."
106 Ibid., 26.
107 Ibid., "l'Œdipe—nom propre devenue nom commun—est un idéogramme qui permet de parler le langage sans langage de l'inconscient."
108 Ibid., "on ne sait le parler en lettres indéfinies quant à leur contenu ou à leurs relations possibles, on ne sait qu'en dessiner des tableaux synthétiques, des images surchargées."
109 Ibid., "Dès lors, plus une forme devient symbolique, plus il est difficile de la penser formellement. L'archétype est un maximum de surcharge signifiante, qu'il soit dieu, héros ou élément."
110 Ibid., 24, "Et, par un court-circuit aveuglants, cet ensemble de choix est à la fois désigne (selon un diagramme en chiasme) comme l'originel des modèles scientifiques clairs, dans une psychanalyse de la connaissance objective et l'originel des archétypes symboliques cultures, dans une psychanalyse de l'imagination matérielle signifiante." [And, by a blinding short circuit, this ensemble of choice is at once designated (by a diagram in chiasm) as the original to scientific and clear models, in a psychoanalysis of objective consciousness and the original to symbolic archetypes, within a psychoanalysis of a material, significant, imagination.] He

comments in footnote 5 that this is the reason why Bachelard, like Baudelaire, never spoke of artificial dreams entitles similar to something like *Haschisch et les rêves*, or *Le Betel et les songes* … opium or mescaline, he notes, are bodies of a non-mythical chemistry, of a non-archetypical chemistry: suddenly alchemy would correspond to true dreams, to a true (and actual) chemistry that corresponds to false images of the sort one will find in Sartre.

111 Ibid., 24.

112 Ibid.

113 Ibid., 25, "Or, pour passer du symbolisme au formalisme, c'est-à-dire du modèle comme fin de la méthode au modèle comme problème, il faut vérifier que la variation est épuisée, des ensembles où l'on peut choisir des archétypes symboliques."

114 Ibid., "C'est en ce sens précis que nous désignons Bachelard comme le dernier des symbolistes: en effet, l'ensemble ou il choisit ses archétypes est le tout de la nature, sans extension imaginable, et l'originel de la nature, sans précession imaginable." [It is in this sense that we call Bachelard the last symbolist: effectively, the ensemble from which he chooses his archetypes is the nature at large, without imaginable extension, and the originality of nature, without anything imaginable that would precede it.]

115 Cf. Michel Serres, "Le Concept de Nature," *Bulletin de l'Académie nationale de médecine* 186, no. 9 (2002): 1683–8; regarding Bruno Latour's recent work, cf. Bruno Latour, *An Inquiry into Modes of Existence: An Anthropology of the Moderns* (Cambridge: Harvard University Press, 2013).

116 Serres, "Structure et importation," 28–31 (footnote 7), "l'histoire est la poche de néguentropie dans l'entropie culturelle."

117 Ibid., "Remonter à rebours une chaine d'information pour éviter les pertes successives de cette dernière définit aussi ce qu'on pourrait appeler le doute historique. Aller à la source, idéale de l'historien et du critique, implique une réciproque à laquelle nul, à ma connaissance, n'a porté une attention suffisante: c'est qu'un contenu historique, par exemple une idée (pour ce qui est de l'histoire de la philosophie) se perd, s'affaiblit, retombe et se mélange. Le vecteur chronologique de l'histoire est porteur de la désagrégation progressive de l'idée. Cette désagrégation n'est pas un oubli pur et simple (comment définir cet oubli?), mais simplement un affaiblissement continue de l' idée par communications successives. L'histoire des idées est ce jeu du téléphone qui donne à la sortie une information d'autant plus déformée que la chaine de communication a été longue."

118 Hénaff, *Price of Truth*, 10.

119 Serres, "Structure et importation," 34.

120 Ibid., 28, "On voir par là ce que nous appelons importation. Un concept méthodique étant défini en précision et claire dans un domaine déterminé et y ayant réussi (une méthode ne peut et ne doit être jugée qu'à ses fruits), on l'essai à l'envie, dans d'autres domaines du savoir, de la critique, etc."

121 Ibid., "Pour rester claire et précis, il suffit d'éviter toute déviation et toute ambiguïté lorsqu'on importe l'idée de structure des théories savantes en général au champ de la critique culturelle."

122 Ibid., "En Algèbre, par exemple, cette idée est dénuée de tout mystère."

123 Ibid., 30. Here Serres elaborates in footnote 8 on such a notion of structure as machine, as a structure that functions in manner analogue to a particular system that is being analysed—this indeed is important to understand what Serres means with his recurring topos of "a parallelism in structure." We will come back to this.

124 Ibid., 31, "En effet, on ne peut importer librement que des concepts hautement formalises. C'est pourquoi le nouveau concept de structure est assez librement importable. Parce qu'il est formel."

125 Ibid., 31, "A ceux qui trouveraient scandaleux d'opérer une telle réduction, nous signalons qu'on a déjà proposé des machines qui fonctionnent comme le système de Darwin. L'idée n'est pas nouvelle, et elle ne parait ignominieuse qu'à ceux qui méprisent la machine pour ignorer ce qu'elle est, peut-être, et doit être. Comme il est instructive et intéressant de voir les philosophes n'accorder attention à la technologie que si elle n'est pas postérieure au préhistorique!" ["It is very telling," he continues, "that philosophers tend to pay attention to technics only when it is of prehistoric character."]

126 Of course, in the history of architecture theory there are many moments when this is treated differently: but these are those moments when one order is to become universal.

127 Cf. *Statues*, especially the beginning, which recounts the foundation of a society in a space shuttle, by burying a dead dog without having an earth that could absorb the body, a foundation, which means, hence, that the bodies of the dead always accompany the founded society through space. And cf. also the end of the book, which describes how Hermes, the statue of mobility, finally joins in an embrace with Hestia, the statue of immobility, and thereby acknowledges that he cannot do without the pursuit of beauty.

128 Serres, "Structure et importation," 28, "une méthode ne peut et ne doit être jugée qu'à ses fruits."

129 Ibid., 34–5, "Donner aux figures de cet autre monde dionysiaque des signification épaisse, compactes et obscures où se projettent l'âme humaine, son affectivité, son destin … Mais, outre qu'elles sont des symboles de l'histoire, ne seraient-elles pas dans leur ultime surcharge, dans leur dernière détermination, des modèle signifiants de structure transparente, de l'intellect, et de la science?"

8 Coda: Architecture in the Meteora

1 Serres, *Geometry*, x, Kindle Edition: position. 52.

2 Serres, *The Incandescent*, 41–2.

3 Cf. the recent study by Malcolm Wilson, *Structure and Method in Aristotles' Meteorologia. A More Disorderly Nature* (New York: Cambridge University Press, 2013).

4 A nice introduction is provided by Craig Martin, *Renaissance Meteorology. Pomponazzi to Descartes* (Baltimore: John Hopkins University Press, 2011).

5 Michel Serres, "Sciences et société," conference given on May 17, 2017 at the University of Strasbourg, available online: https://www.youtube.com/watch?v=FDywhSXVjgw.

6 Serres, *The Incandescent*, 75–6.

7 Serres, *Le Système de Leibniz et ses modèles mathématiques*, Kindle edition: position 162, my own translation from French: "une ordonnance potentielle qui se laisse toujours entrevoir et qui sans cesse se refuse, l' idée vague d' une cohérence perçu mille fois en vue cavalière et qui dérobe son géométral, la sensation de progresser dans un labyrinthe dont il tiendrait le fil sans en avoir la carte. Perspectives offertes, points the vue multipliés, possibilités infiniment itérées: il ne parait jamais qu'on puisse parvenir aux limites exhaustives d'un plan synoptique, étalé, complet, actuel."

8   Ibid., position 524.

9   Ibid., position 487, "par 'infinité d'infinités infiniment répliquée."

10   Ibid., position 483.

11   Ibid., position 493, "des lois de liaison un-multiple, fini-infini, qui valent de manière multivalente pour la perception, la liberté, la connaissance, la création, le souvenir, etc., et qui sont à l'œuvre *aussi* dans le modèle mathématique."

12   Ibid., position 580, "Il n'y a pas là de cause à effet, il y a parallélisme de structure, c'est pourquoi nous parlons de modèle mathématique, et seulement de modèle."

13   Serres, *The Parasite*, 43.

9 Instead of a Conclusion: The Static Tripod

1   Serres, *Statues*, 162–70.

Bibliography

Bachelard, G. (1934), *Le nouvel ésprit scientifique*, Paris: Alcan.

Bachelard, G. (1940), *La philosophie du non: essai d'une philosophie du nouvel esprit scientifique*, Paris: Presse Universitaire France.

Bahr, H.-D. (1983), *Über den Umgang mit Maschinen*, Tübingen: Konkursbuch.

Barad, K. (2007), *Meeting the Universe Halfways: Quantum Entanglements of Matter and Meaning*, Durham and London: Duke University Press.

Barthes, R. ([1953] 1972), *Le Degré zéro de l'écriture*, Paris: Seuil.

Barthes, R. ([1964] 1983), "The Eiffel Tower," in Susan Sontag (ed.), *A Barthes Reader*, 236–50, New York: Hill and Wang.

Blanché, R. ([1955] 1980), *L'Axiomatique*, Paris: Presses Universitaires de France.

Blanchot, M. (1959), *Le Livre à venir*, Paris: Gallimard.

Brillouin, L. ([1956] 2013), *Science and Information Theory*, New York: Academic Press. (Kindle edition).

Brook, A., and Jennifer McRobert (1998), "Kant's Attack on the Amphiboly of the Concepts of Reflection," *Theory of Knowledge*, August 1998, available online: https://www.bu.edu/wcp/Papers/TKno/TKnoBroo.htm (accessed January 12, 2018).

Bühlmann, V. (2015), "Foederae Naturae. Eine Annäherung zwischen Jean-Luc Nancy's Begriff der Exscription und Michel Serres' Begriff der transzendentalen Allgemeinheit von Information," unpublished manuscript, available online: https://www.academia.edu/31872221/Foederae_Naturae._Eine_Annäherung_zwischen_Jean-Luc_Nancy_s_Begriff_der_Exscription_und_Michel_Serres_Begriff_der_transzendentalen_Allgemeinheit_von_Information (accessed January 7, 2018).

Bühlmann, V. (2015), "Ichnography—The Nude and Its Model. The Alphabetic Absolute and Storytelling in the Grammatical Case of the Cryptographic Locative," in V. Bühlmann et al. (eds.), *Coding as Literacy, Metalithikum IV*, 276–350, Basel/Vienna: Birkhäuser.

Bühlmann, V. (2017), "Generic Mediality: On the Role of Ciphers and Vicarious Symbols in an Extended Sense of Code-based 'Alphabeticity'," in Rosi Braidotti and Rick Dolphijn (eds.), *Philosophy after Nature*, 31–54, London: Rowman and Littlefield.

Bühlmann, V. (2018), "Architectonic Dispositions," in Rosi Braidotti and Maria Hlavajova (eds.), *Posthuman Glossary*, 55–9, London: Bloomsbury.

Bühlmann, V. (2018), "Cosmoliteracy and Anthropography: An Essay on Michel Serres' Book *The Natural Contract*," in Rick Dolphijn (ed.), *Michel Serres and the Crisis of the Contemporary*, 31–50, London: Bloomsbury.

Bühlmann, V. (2018), "Equation," in Rosi Braidotti and Maria Hlavajova (eds.), *Posthuman Glossary*, 33–8, London: Bloomsbury.

Bühlmann, V. (2018), "Invariance," in Rosi Braidotti and Maria Hlavajova (eds.), *Posthuman Glossary*, 212–16, London: Bloomsbury.

Bühlmann, V. (2018), "Maxwell's Demon (Non-Anthropocentric Cognition)," in Rosi Braidotti and Maria Hlavajova (eds.), *Posthuman Glossary*, 247–51, London: Bloomsbury.

Bühlmann, V. (2018), "Negentropy," in Rosi Braidotti and Maria Hlavajova (eds.), *Posthuman Glossary*, 273–77, London: Bloomsbury.

Boole, G. ([1853] 2009), *An Investigation into the Laws of Thought on Which Are Founded the Mathematical Theories of Logic and Probabilities*, Cambridge: Cambridge University Press.

Crahay, A. (1988), *Michel Serres. La Mutation du Cogito. Genèse du Transcendental Objectif*, Bruxelles: de Boeck.

Deleuze, G. (1968), *Difference and Repetition*, Paris: Presses Universitaires de France.

Deleuze, G., and Guattari, F. (1980), *Mille Plateaux*, Paris: Les Éditions de Minuit.

Derrida, J. (1974), "White Mythology: Metaphor in the Text of Philosophy," in F. C. T. Moore (trans.), *New Literary History*, 6 (1), On Metaphor, 5–74, John Hopkins University Press, Autumn.

Detienne, M. (1996), *The Masters of Truth in Archaic Greece*, New York: Zone Books.

Euclid ([1908] 1956), *Euclid's Elements*, trans. Thomas L. Heath, New York: Dover.

Feynman, R. ([1986] 2014), *QED: The Strange Theory of Light and Matter*, Princeton: Princeton University Press.

Feynman, R. ([1964] 2017), *The Character of Physical Law*, Cambridge: MIT Press.

Fischer, G. (2009), *Vitruv NEU, oder was ist Architektur?* Basel: Birkhäuser.

Gaither, C. and Cavazos-Gaither A., eds. (2012), *Gaither's Dictionary of Scientific Quotations*, New York: Springer.

Girard J.-Y. (2014), "Three lectures on 'certaintainy' at the Centre Henri Poincaré," available online: https://www.youtube.com/watch?v=f7sT0J74pHI (accessed January 2, 2018).

Gleick, J. (2011), *Information: A History, a Theory, a Flood*, London: Harper Collins.

Grätzer, G. ([1968]/2008), *Universal Algebra*, New York: Springer.

Hailperin, T. (1981), "Boole's Algebra Isn't Boolean Algebra," *Mathematics Magazine*, 54 (4): 172–84.

Hegel, G. F. W. ([1872/1900] 1991), *Lectures on the Philosophy of History*, trans. J. Sibree, New York: Prometheus.

Hénaff, M. (2010), *The Price of Truth: Gift, Money, and Philosophy*, trans. Jean-Louis Morhange, Stanford: Standford University Press.

Jacobs, W. (2010), "Das Absolute," in H. J. Sandkühler (ed.), *Enzyklopädie Philosophie*, 13–17, Hamburg: Felix Meiner Verlag.

Kant, I. ([1781] 1998), "'Amphiboly of Concepts of Reflection,' Appendix to 'The Transcendental Analytic,'" in I. Kant (ed.), *Critique of Pure Reason*, trans. Paul Guyer, Allen W. Wood, 366–83, Cambridge: Cambridge University Press.

Kant, I. ([1800] 1966), *Logik, Ein Handbuch zu Vorlesungen (1800)*, ed. Friedrich Nicolovius, Königsberg: Akademie-Ausgabe, Bd. 9, 1–150, A 17.

Kesten, H. (1982), *Percolation Theory for Mathematicians*, Basel: Springer.

Kosmann-Schwarzback, Y. (2011), *The Noether Theorems. Invariance and Conservation Laws in the Twentieth Century*, trans. Bertram E. Schwarzbach, New York: Springer.

Lambert, G. (2017), *Philosophy after Friendship: Deleuze's Conceptual Personae*, Minneapolis: University of Minnesota Press.

Levy-Leblond, J. M. and Balibar, F. (1990), *Quantics: Rudiments of Quantum Physics*, Amsterdam: North Holland.

Lyotard, J.-F. ([1979] 1984), *The Postmodern Condition: A Report on Knowledge*, trans. G. Bennington and R. Bowlby, Minneapolis: University of Minnesota Press, 191–204.

Longo, G. (2015), "Synthetic Philosophy of Mathematics and Natural Sciences," CNRS, Collège de France et École Normale Supérieure, Paris, available online: https://www.

di.ens.fr/users/longo/files/PhilosophyAndCognition/Review-Zalamea-Grothendieck.
   pdf (accessed October 2, 2019).

Marinčić, Nikola (2017), *Towards Communication in CAAD. Spectral Characterisation and
   Modelling with Conjugate Symbolic Domains*, PhD thesis, Zurich: ETH Zurich.

Massa Esteve, M. R. (2008), "Symbolic Language in Early Modern Mathematics: The
   Algebra of Pierre Hérigone (1580–1643)," *Historia Mathematica*, 35 (4): 285–301.

Monod, J. ([1970] 1971), *Chance and Necessity. An Essay on the Natural Philosophy of
   Modern Biology*, New York: Vintage.

Odifreddi, P., and Cooper, S. B. (2012), "Recursive Functions," in Edward N. Zalta (ed.),
   The Stanford Encyclopedia of Philosophy (Fall 2012), available online: http://plato.
   stanford.edu/archives/fall2012/entries/recursive-functions (accessed January 12, 2018).

Orton, J. (2004), *The Story of Semiconductors*, Oxford: Oxford University Press.

Peckhaus, V. (2006), "Die Aktualität der Logik als Organon," in G. Abel (ed.),
   Kreativität: XX. Deutscher Kongress für Philosophie, 26–30 September 2005 an der
   Technischen Universität Berlin: Kolloquienbeiträge, Hamburg: Felix Meiner Verlag.

Reeves, H. et al. ([1996] 2011), *Origins. Cosmos, Earth and Mankind*, New York: Arcade.

Ritter, J. ([1989] 1995), "Babylon," in Serres (ed.), *History of Scientific Thought*, 17–43,
   Cambridge: Blackwell.

Ritter, J. ([1989] 1995), "Measure for Measure," in Serres (ed.), *History of Scientific
   Thought*, 44–72, Cambridge: Blackwell.

Schrödinger, E. (1944), *What Is Life? The Physical Aspect of the Living Cell*,
   Cambride: Cambridge University Press.

Schubring, G. (2016), "From Pebbles to Digital Signs: The Joint Origin of Signs for
   Numbers and for Scripture, Their Intercultural Standardization and Their Renewed
   Conjunction in the Digital Era," in V. Bühlmann and L. Hovestadt (eds.), *Symbolizing
   Existence, Metalithikum III*, 114–26, Vienna: Birkhäuser.

Serres, J.-F., and M. (2015), *Solitude. Dialogue sur l'engagement*, Paris: Le Pommier.

Serres, M. (1968), *Hermes I, La Communication*, Paris: Éditions de Minuit.

Serres, M. (1968), "La querelle des Anciens et des Modernes," in M. Serres (ed.), *Hermes I,
   La Communication*, 46–77, Paris: Éditions de Minuit.

Serres, M. (1968), "Le réseau de communication: Pénélope," in M. Serres (ed.), *Hermes I,
   La Communication*, 11–35, Paris: Éditions de Minuit.

Serres, M. (1968), *Le Système de Leibniz et ses modèles mathématiques*, Paris: Presses
   universitaires de France.

Serres, M. (1968), "Les anamnèses mathématiques," in M. Serres, *Hermes I, La
   communication*, 78–112, Paris: Les Editions de Minuit.

Serres, M. (1968), "Structure et importation: des mathématiques aux mythes," in Serres,
   M., *Hermes I, La Communication*, 21–35, Paris: Les Éditions de Minuit.

Serres, M. (1972), "Ce que Thalès a vu au pied des Pyramides," in M. Serres, *Hermes II,
   L'Interférence*, 163–80, Paris: Éditions de Minuit.

Serres, M. (1972), "Ce qui est écrit sur la table rase," in M. Serres, *Hermes II, L'Interférence*,
   67–126, Paris: Les Éditions de Minuit.

Serres, M. (1972), *Hermes II, L'Interférence*, Paris: Éditions de Minuit.

Serres, M. (1972), "L'Interférence théorique: tabulation et complexité," in M. Serres,
   Hermes II, L'Interférence, xx, Paris: Éditions de Minuit.

Serres, M. (1972), "Mathématisation de l'empirisme," in M. Serres, *Hermes II,
   L'Interférence*, 195–200, Paris: Les Éditions de Minuit.

Serres, M. (1974), *Hermes III, La Traduction*, Paris: Les Éditions de Minuit.

Serres, M. (1974), "Trahison: la thanatocratie," in M. Serres, *Hermes III, La Traduction*, 73–106, Paris: Les Éditions de Minuit.

Serres, M. (1974), "Vie, information, deuxième principe," in M. Serres, *La Traduction, Hermes III*, 43–72, Paris: Les Éditions de Minuit.

Serres, M. (1975), *Feux et signaux de brume. Zola*, Paris: Grasset.

Serres, M. (1977), "Moteurs, suivi de régression," in M. Serres, *Hermes IV, La Distribution*, 41–86, Paris: Les Éditions de Minuit.

Serres, M. (1980), "Espace et temps," in M. Serres, *Hermes V, Le passage du nord-ouest*, 67–83, Paris: Éditions de Minuit.

Serres, M. (1980), *Hermes V, Le passage du nord-ouest*, Paris: Éditions de Minuit.

Serres, M. (1980), "Histoire: L'Univers et le lieu. Obstruction," in M. Serres, *Hermes V, Le passage du nord-ouest*, 84–92, Paris: Éditions de Minuit.

Serres, M. (1980), "Solides, Fluides, Flames," in M. Serres, *Hermes V, Le Passage Nord-Ouest*, Paris: Les Éditions de Minuit, xx.

Serres M. ([1980] 1982), *The Parasite*, trans. L. Schehr and R. Laurence, Baltimore: John Hopkins University Press.

Serres, M. (1983), "Noise," trans. Lawrence R. Schehr, *SubStance*, 12 (3): 48–60.

Serres, M. (1987), *L' Hermaphrodite*, Paris: Flammarion.

Serres, M. (1990), *Le Contrat Naturale*, Paris: François Bourin.

Serres, M. (1991), *Le Tiers-Instruit*, Paris: François Bourin.

Serres, M. (1993), *La Légende des Anges*, Paris: Flammarion.

Serres, M. ([1982] 1995), *Genesis*, trans. Geneviève James and James Nielson, Ann Arbor: University of Michigan Press.

Serres, M. ([1989] 1995), "Gnomon: The Beginnings of Geometry in Greece," in M. Serres (ed.), *A History of Scientific Thought*, 73–123, Cambridge: Blackwell.

Serres, M. ed. ([1989] 1995), *A History of Scientific Thought*, Cambridge: Blackwell.

Serres, M. ([1990] 1995), *The Natural Contract*, trans. Elizabeth MacArthur and William Paulson, Ann Arbor: University of Michigan Press.

Serres, M. ([1994] 1996), *Atlas*, Paris: Flammarion.

Serres, M. (1997), *Les Nouvelles du Monde*, Paris: Flammarion.

Serres M. ([1977] 2000), *The Birth of Physics in the Text of Lucretius*, trans. Jack Hawkes, Manchester: Clinamen Press.

Serres, M. (2001), *Hominescence*, Paris: Le Pommier.

Serres, M. (2002), "Fraternal Confession," trans. Kris Pender, unpublished, available online: https://www.academia.edu/11074066/Michel_Serres_-_Fraternal_Confession (accessed January 12, 2018).

Serres, M. (2002), "La confession fraternelle," *Empan* (48): 11–16, available online: https://www.cairn.info/revue-empan-2002-4-page-11.htm (accessed January 12, 2018).

Serres, M. (2003), *L'Incandescent*, Paris: Le Pommier.

Serres, M. (2004), *Rameux*, Paris: Le Pommier.

Serres, M. (2006), "Revisiting the Natural Contract," in Arthur and Marilouise Kroker (eds.), *1000 Days of Theory*, available online: http://ctheory.net/ctheory_wp/revisiting-the-natural-contract/ (accessed January 12, 2018).

Serres, M. (2007), "Les nouvelles technologies: révolution culturelle et cognitive," Lecture by Michel Serres for the 40th anniversary of l'INRIA, Paris, available online: https://interstices.info/jcms/c_33030/les-nouvelles-technologies-revolution-culturelle-et-cognitive (accessed January 12, 2018).

Serres, M. ([1998] 2008), *The Five Senses: A Philosophy of Mingled Bodies*, trans. Margaret Sankey and Peter Cowley, Manchester: Continuum.

Serres, M. (2008), *Le mal propre. Polluer pour s'approprier?* Paris: Le Pommier.

Serres, M. (2010), "Interview de Michel Serres sur l'autorité," APEL the Association of Parents of School Kids in France, in the context of a Conference entitled Autoriser l'autorité in Montpellier 2010, available online: https://www.youtube.com/watch?v=GeJgTmZ9EBc (accessed January 2, 2018).

Serres, M. (2011), "Sacré vs Spiritualité," Lecture in Rennes: créa, Éspace des Science, available online: https://www.youtube.com/watch?v=wrMZCHYWsKU (accessed January 2, 2018).

Serres, M. (2012), "Les Sciences et la Langue," Rencontres Science et Humanisme 2012, Ajaccio, available online: https://www.youtube.com/watch?v=JV8E5w3nClk (accessed December 27, 2017).

Serres, M. (2014), "Le Grand Fétiche ou les metamorphoses du religieux," in M. Serres, *Pantopie: de Hermes à Petite Poucette. Entretiens avec Martin Legros et Sven Ortoli*, 263–96, Paris: Le Pommier.

Serres, M. (2014), "Le Thanatocrate ou le pouvoir de la mort," in M. Serres, *Pantopie: de Hermes à Petite Poucette. Entretiens avec Martin Legros et Sven Ortoli*, 161–84, Paris: Le Pommier.

Serres, M. (2014), "Pantope ou penser en inventant des personnages," in M. Serres, *Pantopie: de Hermes à Petite Poucette. Entretiens avec Martin Legros et Sven Ortoli*, 69–114, Paris: Le Pommier.

Serres, M. (2014), *Pantopie: de Hermes à Petite Poucette. Entretiens avec Martin Legros et Sven Ortoli*, Paris: Le Pommier.

Serres, M. (December 18, 2014), "Yeux," lecture at the Institute Catholique de Paris, available online: https://www.youtube.com/watch?v=1qFdYgjWg9s (accessed December 27, 2017).

Serres, M. (2015), *Du Bonheur aujourd'hui*, Paris: Le Pommier.

Serres, M. (2015), "La culture scientifique et le numérique," Conférence donnée dans le cadre du 33ème congrès de l'Amcsti—Association des musées et centres pour le développement de la culture scientifique et industrielle—June 23, 2015 à Chambéry, available online: https://www.youtube.com/watch?v=_eyKWOonh1c&t=418s (accessed January 12, 2018).

Serres, M. (2015), "La culture scientifique et le numérique," Lecture delivered at the 33rd Congress by Amcsti—Association des musées et centres pour le développement de la culture scientifique et industrielle—June 23, 2015 à Chambéry, available online: https://www.youtube.com/watch?v=_eyKWOonh1c&t=418s (accessed January 12, 2018).

Serres, M. (2015), "Le don," in *Du Bonheur aujourd'hui*, 40–5, Paris: Le Pommier.

Serres, M. (2015), *Le gaucher boiteux: puissance de la pensée*, Paris: Le Pommier.

Serres, M. ([1983] 2015), *Rome: The First Book of Foundations*, trans. Randolph Burks, London: Bloomsbury.

Serres, M. ([1987] 2015), *Statues: The Second Book of Foundations*, trans. Randolph Burke, London: Bloomsbury.

Serres, M. (2016), *Darwin, Bonaparte et le Samaritain, Une philosophie de l'histoire*, Paris: Le Pommier.

Serres, M. ([1993] 2017), *Geometry: The Third Book of Foundations*, trans. Randolph Burks, London: Bloomsbury.

Serres, M. (2017), "Information and Thinking," in Rick Dolphjin and Rosi Braidotti (eds.), *Philosophy after Nature*, 13–20, London and New York: Rowman and Littlefield.

Serres, M. (2017), "Le problème de la violence a été au cœur de ce que j'ai produit depuis 70 livres," Masterclasse avec Michel Serres, France Culture August 23, 2017, available

online: https://www.franceculture.fr/emissions/les-masterclasses/michel-serres-le-
  probleme-de-la-violence-ete-au-coeur-de-ce-que-jai (accessed December 27, 2017).
Serres, M. ([2003] 2018), *The Incandescent*, trans. Randolph Burks,
  London: Bloomsbury Press.
Shannon, C. (1948), "A Mathematical Theory of Communication," *Bell System Technical
  Journal*, 27 (3): 379–423.
Simondon, G. (1983), "Save the Technical Object," interview by Anita Kechickian (Esprit
  76:147–52. 04/1983), trans. Andrew Illiadis (unpublished), available online: https://
  www.scribd.com/document/166278390/Simondon-Gilbert-Save-the-Technical-Object
  (accessed January 12, 2018).
Sloterdijk, P. (2001), *Tau von den Bermudas. Uber einige Regimes der Einbildungskraft*,
  Frankfurt am Main: Suhrkamp.
Spinoza, B. ([1677] 1937), *Ethics, Demonstrated in Geometrical Order*, trans. William Hale
  White, Oxford: Oxford University Press.
Stengers, I. ([2003] 2010), *Cosmopolitics I*, Minneapolis: University of Minnesota Press.
Stenlund, S. (2014), *The Origin of Symbolic Mathematics and the End of the Science of
  Quantity*, Uppsala: Uppsala University Press.
Turing, A. (1937), "On Computable Numbers, with an Application to the
  Entscheidungsproblem," Proceedings of the London Mathematical Society, 42: 230–65.
Turing, A. M. (1952), "The Chemical Basis of Morphogenesis," *Philosophical Transactions
  of the Royal Society of London*, series B, 237 (641): 37–72.
Vuillemin, J. (1962), *La Philosophie de l'Algèbre, Tome I: Recherches sur quelques concepts et
  méthodes de l'Algèbre moderne*, Paris: Presse Universitaire.
Vuillemin, J. (1996), *Necessity or Contingency: The Master Argument*, Chicago: University
  of Chicago Press.
Vuillemin, J. (2009), *What Are Philosophical Systems?* Cambridge: Cambridge
  University Press.
Watson, D. L. (1930), "Entropy and Organization," *Science*, 72 (1861) (August): 220–2.
Weyl, H. (1939), "Invariants," *Duke Mathematical Journal*, 5 (3): 489–502.
Weyl, H. ([1918] 1994), *The Continuum: A Critical Examination of the Foundation of
  Analysis*, London: Dover.
Whitehead, A. N. (1898), *A Treatise on Universal Algebra*, Cambridge: Cambridge
  University Press.
Whitehead, A. N. (1911), *Introduction to Mathematics*, New York: Henry Holt.
Wiener, N. ([1948] 1985), *Cybernetics: Or Control and Communication in the Animal and
  the Machine*, Cambridge: MIT Press.
Zafiris, E. (forthcoming), *Natural Communication. The Obstacle-Embracing Art of Abstract
  Gnomonics*, Basel/Vienna: Birkhäuser.
Zafiris, E. (2015), "The Nature of Local/Global Distinctions, Group Actions, and Phases: A
  Sheaf-Theoretic Approach to Quantum Geometric Spectra," in Vera Bühlmann et al.
  (eds.), 172–86, *Coding as Literacy, Metalithikum IV*, Vienna: Birkhäuser.
Zalamea, F. (2012), *Synthetic Philosophy of Contemporary Mathematics*, London:
  Urbanomic.

Index

Printed in Great Britain
by Amazon

50901677R00150